The Analysis of Geographical Data

The Analysis of Geographical Data

W. H. Theakstone
Lecturer in Geography in the University of Manchester

and

C. Harrison
Lecturer in Geography in the University of Salford

HEINEMANN EDUCATIONAL
BOOKS·LONDON

Heinemann Educational Books Ltd

LONDON EDINBURGH MELBOURNE TORONTO
AUCKLAND SINGAPORE JOHANNESBURG
HONG KONG NAIROBI IBADAN NEW DELHI

Cased edition ISBN 0 435 346903
Paperback edition ISBN 0 435 346911

Published by Heinemann Educational Books Ltd
48 Charles Street, London W1X 8AH   .
Printed in Great Britain by Butler and Tanner Ltd
Frome and London

Contents

72-9097

Acknowledgements

We thank the following people for their help in the production of this book: Mr Peter Dicken, Lecturer in Geography in the University of Manchester, for permission to use some of his work on the Manchester clothing industry and for valuable suggestions in the initial stages of the production of the volume; Miss Mary Hamilton, for permission to use some of her work on the island of Rousay; Mrs Pauline Stuart of the University of Salford for her patience in typing several drafts of the text; and Mr Reginald Oliver of the University of Salford for preparing the maps and diagrams.

We also thank the Scottish Record Office and H.M.S.O. for permission to reproduce statistical information.

Preface

In recent years the study of geography has become more quantitative in approach and this trend has been exemplified by the increasing attention paid to statistical techniques in first degree courses in geography in British universities. There is a need, therefore, for an introductory book about the statistical analysis of the sorts of data which geographers handle, data broadly concerned with spatial distribution.

In this book an attempt is made to place the emphasis on the types of problem which are faced by the geographer in particular, although many of these are similar to those which other scientists face. Consequently, the book is not intended to be a comprehensive text on statistical methods with geographical examples. It is an elementary *geographical* text, which it is hoped will be of use to students beginning statistics courses at universities and colleges of education, and to those writing dissertations and project reports, who need basic statistical techniques in order to use most effectively the data which they have collected. With the inevitable spread to the schools of the use of statistical techniques in geography, this book should also be of use to sixth formers conducting local studies and to their teachers, many of whom qualified before the value of statistical methods in dealing with geographical problems was fully appreciated.

Because this book is intended as a first text, its scope is limited to the range of work where the necessary calculations can be carried out without the aid of a computer, that is, with logarithm tables, a slide rule, or preferably a desk calculator. Throughout the book it is assumed that the reader has no more than a rudimentary acquaintance with mathematics.

The book is built around three basic sets of information, drawn from some of the varied subjects which occupy much of the geographer's time, namely patterns of agriculture, industry and population. Our object in writing this text has been to help geographers to use statistical methods to clarify the nature of a problem by stating it more clearly, rather than to provide an answer to it.

We hope that our method of approaching the use of statistical methods through particular geographical situations, rather than by taking the techniques as the starting point, will overcome some of the natural disinclination of students of geography to tackle numerical problems. If it does, then it will have been worth while.

W. H. Theakstone
Manchester and Aarhus

C. Harrison
Salford
1970

List of Figures

List of Tables

1. Introduction

Three contrasting sets of data are considered in this book. The authors believe that it is more worth while to deal at length with each of a small number of situations than to cite many individual unrelated sets of data, to each of which reference is made only once. The latter approach inadvertently may give the impression that there is only one way of dealing with a particular sort of information or problem. Any of the three sets of data used here could form the basic raw material for an elementary geographical project or dissertation. The techniques used therefore are those which are appropriate to this level of investigation and scale of operation.

In some projects, the geographer collects data in the field; in others he obtains data from existing sources. The material used in this book covers both these situations, as well as that in which some data are already available, but more have to be collected by the 'field worker'. The data refer to aspects of agriculture, industry and population; no example from the field of physical geography is used. This is not because physical geography is regarded as of less importance, but because the use of physical data may mask some of the problems which are not apparent when information is complete and accurate, standardised and in convenient units which are easily compared. Much of the material of physical geography, especially that dealt with by climatologists, is of this type. A number of issues which arise because of practical problems in data collection are most conveniently discussed in terms of human geography, in which fields they are most likely to be encountered. Problems in physical geography, of course, may be tackled with the aid of the techniques applied in the following chapters.

The island of Rousay (Figure 1.1) is one of the North Isles of the Orkney archipelago. Eighteen square miles in area, it rises to 800 feet above sea level, and a central depression separates hills close to its northern and southern shores. Boulder clay and peat cover parts of the flagstones which make up most of the area. The North Atlantic

1

Drift maintains temperatures which are high for the latitude, and hard frosts are almost unknown. The high latitude (59°N) results in short winter days but long summer ones. The annual rainfall is only about 32 inches, but rain falls on 225–250 days of the year. The frequent rainfall and long summer daylight favour vegetative growth, but prolonged cloudiness inhibits the ripening of cereals. Grass

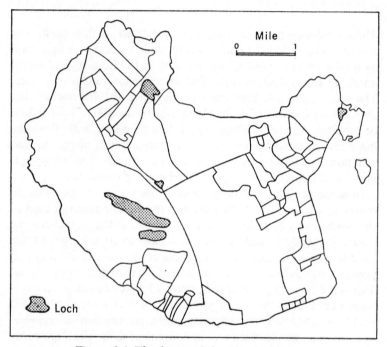

Figure 1.1 The farms of the island of Rousay
(*Some consist of several separate areas*)

is therefore the principal crop. The peat-covered areas are used mainly for rough grazing, the flagstones and till providing more arable land. In 1841, the population of Rousay was nearly 1000; since then, however, decline has been almost continuous. In 1966, when a study of the island was made, there were 232 inhabitants.

As detailed information was not otherwise available in 1966, data about Rousay had to be collected in the field. Since there were only 44 farms, material could be obtained for all of them. The data, presented in Appendix A, concern the areas of farms, the proportion of arable land, and the number of livestock. The collection of standard information about all the geographical units (in this case, farms) within an area is possible only in rare instances, for time usually restricts to small areas or islands, such as Rousay, the gathering of comprehensive data. There are great advantages in having complete information: for instance, we can state without qualification that 21·7 per cent of Rousay is arable land, the percentage being an absolute value, not an estimate.

Amongst the industries of the Manchester conurbation is the clothing trade. Some basic information about the 826 geographical units (in this case, firms) involved is readily available; for instance, the names and addresses of the firms can be obtained from the classified telephone directory, a very useful source of material for the geographer in some of his studies. Amongst other groups of firms, the directory for the Manchester area lists those in business as Blouse Makers, Glove Manufacturers, Shirt and Collar Makers, and Skirt Manufacturers. However, in order to study the Manchester clothing industry, much more information than a mere name and address is needed for each firm. Because the total number of firms is relatively large, time would not permit a visit to each in order to discover, for example, exactly what it manufactured and how many workers were employed. It was necessary, therefore, to select a sample of firms representative of them all and to visit these alone. Thus, a figure for the proportion of firms manufacturing one type of clothing cannot have absolute validity, but is an estimate based on the sample studied. The data relating to the clothing industry of the Manchester conurbation which are considered in this book are presented in Appendix B.

Details of the population of the United Kingdom are collected on a regular basis by means of the Census of Population. The amount of available data, which has been partly processed between the times of collection and publication, is very great indeed. Thus, it is possible for the geographer to undertake a study of aspects of population distribution in Lancashire in 1961 without himself having to collect data in the field. The published lists indicate the number of people living in the County Boroughs, Municipal Boroughs, Urban Districts and Rural Districts of Lancashire at the time of the 1961 census. The County Boroughs are large, only Barrow-in-Furness and Bury having fewer than 75,000 inhabitants, and it is with the population of the generally smaller towns, the 95 Municipal Boroughs and Urban

Districts, that we are principally concerned in this book; the relevant data are presented in Appendix C.

The practical problems associated with data collection must be appreciated, and must be borne in mind constantly, when material is being analysed and conclusions are being drawn. Unlimited time is rarely available for geographical studies, so that the methods of data collection used may be influenced by the short period in which the operation has to be carried out. A second problem, more common in human geography though by no means absent from the physical side, concerns the reliability of information. For example, an employer being interviewed about the number of employees in his firm may give the precise number if it is small, but if he employs many workers he may 'round off' the number. It is possible, too, that the person interviewed may not tell the truth—he may wish, for instance, to exaggerate the 'importance' of his firm and so quote a size greater than the real one. Some information, particularly that of a financial nature, may be difficult or impossible to obtain. Sometimes this problem can be overcome by an undertaking that information will be disclosed only in an aggregate form, so that details about individuals are not revealed. Such a course generally is practicable as statistical analysis is concerned mainly with groups of units, rather than with individuals. However, knowledge of precise individual values is important in those stages of calculation which lead to a general statement about the group. If information is available initially only in aggregated form, difficulties arise which, although not insurmountable, are inconvenient.

The nature of geographical problems could in itself be the subject of an entire book, and we do not intend to discuss either this or the philosophy and nature of geography. This book concerns elementary geographical problems which can be expressed in numerical terms. The problems discussed are 'geographical' in any of the commonly held views of the subject, whether it be the study of spatial distributions, areal differentiation or the relationships between the elements of the landscape. The problems with which we deal are:

(a) How can masses of numerical information be described and summarised in simple terms?

(b) How can estimates of the various characteristics of geographical units be made when information about some units is not available, or its collection is impracticable?

(*c*) How can we discover whether the geographical characteristics of one area differ from those of another?

(*d*) How can relationships between geographical phenomena be established?

These problems are both statistical and geographical. Having solved them, using statistical techniques in a geographical context, the true geographical questions emerge, e.g. *Why* is one area different from another? *Why* do relationships exist between phenomena? In other words, attempts to decide which elements are related and how the relationships may be discovered involve the use of techniques which enable us to state clearly the true geographical problem: Why? In this book, we deal with each of the four technical problems in turn, drawing on the sets of data where appropriate, rather than taking each set separately, which would involve much repetition. However, statistical methods are a preliminary to thought, not an abdication of it, and the questions arising from the statistical treatment of the data are themselves discussed in the final part of the volume.

In statistical work, symbols denoting particular measures are of great use. A glossary of all the symbols used in the text is presented at the end of the book. Statistical and logarithm tables and a glossary of terms are also provided.

2. Summarising information

In any geographical study where there is a mass of numerical information, the first problem to arise is how to summarise the information and express it simply and concisely in numerical terms. General statements tend to be more useful than do descriptions of each individual item in a group.

Certain data are of such a type that different parts can be compared and expressed in percentage terms. For example, the 1961 Census of Population for Lancashire gives the population of each County Borough, Municipal Borough, Urban District and Rural District, but the characteristics of the population are not immediately apparent from the list alone. To state that 6·3 per cent of the population live in Rural Districts is much more helpful. Information collected about Rousay in 1966 revealed that the 44 land holdings totalled 11 002 acres, 2387 being given to arable use. At the time of the survey, therefore $\frac{2387}{11\ 002} \times 100$ or 21·7 per cent of the land farmed on the island was arable.

Percentages are proportional comparisons and require that the information be divided into separate groups, each with distinct characteristics. Thus, agricultural land may be classed as either arable or pastoral, the two being distinctly different. No one piece of land could be given to both pastoral and arable use at the same time. A summary of information about land use in terms of percentages, then, is very useful. Merely by listing the percentage of land on Rousay within certain categories—arable, pastoral, waste land, etc—we can provide an immediate picture of the predominant types of land use on the island at a particular time. Consider, however, the farm sizes on Rousay, where the differences between individual units are a matter of degree rather than of kind. To say that 65·9 per cent of the farms are larger than 50 acres is factually correct, but it does not tell us a great deal about the farm sizes; a rapid look at the list of sizes in its untreated form probably would tell us a great deal more. Only if 50 acres was chosen on objective grounds, in that it divided farms into two groups different in some

6

attribute other than that of size, might a percentage expression be useful.

Clearly, in making a summary of farm sizes on Rousay, some sort of 'average' value is helpful. However, several ways of obtaining an 'average' exist, and it is necessary to decide which provides the best impression of the real situation, the distribution by size, or *size frequency distribution*, of the holdings.

The *arithmetic mean size* ($\bar{x}$) is obtained by adding together the sizes of the individual holdings (x_1, x_2, x_3, etc.) and dividing the total by the number of holdings (n). This may be expressed as:

$$\bar{x} = \frac{\Sigma x}{n}$$

where x represents the size of the holding; Σ (sigma) simply indicates that the individual values of x are added together, i.e. $\Sigma x = x_1 + x_2 + x_3 + \ldots + x_n$. The arithmetic mean size of the 44 holdings on Rousay thus is $\frac{11\ 002}{44}$ acres, or 250·04 acres. However, as 36 of the 44 holdings are smaller than this 'average' size, the arithmetic mean value could hardly be described as representative, and the measure has little relevance in this case.

A much better impression of the 'average' size of units in such a situation is given by the *median value*, which is that above and below which there are equal numbers of units. If the total number of units, n, is odd, the median value is that which is $\frac{1}{2}(n + 1)$th in the order. If n is even, the median value is taken as the mean of the two central ones. Thus, in the case of the Rousay holdings, where $n = 44$, the median value is derived from the sizes of the holdings which are *ranked* 22nd and 23rd in the order, that is, the median size is $\frac{76 + 100}{2}$ acres, or 88 acres. Clearly, this 'average' size, above and below which there are 22 farms, is a rather more useful measure than is that represented by the arithmetic mean, since it indicates a size near the mid-point of the total range. The data of Appendix A indicate that two sizes of farm (4 and 40 acres) are more common than any of the others. These *modal sizes*, in which the frequency of units is greatest, are a third sort of average.

Each holding on Rousay may be placed within a single 100 acre class, and the size distribution of the holdings then may be represented by a *histogram*, or bar graph, showing the number of holdings in each class (Figure 2.1). The histogram shows that there are many small holdings and a few very large ones; in other words the distribution is asymmetric, or *skewed*. In this particular case the large holdings provide the 'tail' of the distribution, which is therefore described as *right*, or *positively*, skewed. In a symmetrical distribution, the arithmetic mean and median values coincide but in a skewed distribution they differ (Figure 2.2). Generally speaking, the greater the skewness, the less representative is the arithmetic mean value, and the more useful does the median become.

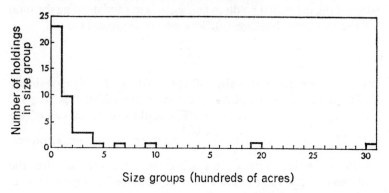

Size groups (hundreds of acres)

Figure 2.1 A histogram of the size of Rousay farms

Clearly, it is meaningless to talk about an 'average' size without qualifying the term and making it more precise. No single measure of the 'average' is the most suitable for all circumstances. Sometimes, the mean value may provide the most useful indication of the *central tendency* of a set of data but in other cases the mode or median may be better. In many cases the different measures can be used together as each supplements the information provided by the others.

Sometimes the data summarised are in grouped form. In some cases this is because the actual values of individual items were not collected, but merely put into appropriate classes, so that only the number of items within each of several classes is known. In other cases, the large number of individual values, even if known, makes the calculation of the mean a long process as all separate values of x have to be summed in order to calculate the mean $\bar{x}$, using the expression $\dfrac{\Sigma x}{n}$. For instance, in determining the mean size of Lancashire's Municipal Boroughs and Urban Districts in 1961, 95 separate values would have to be added together.

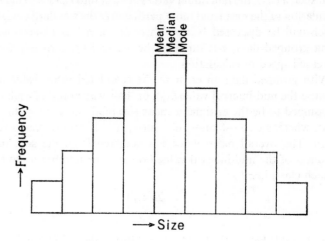

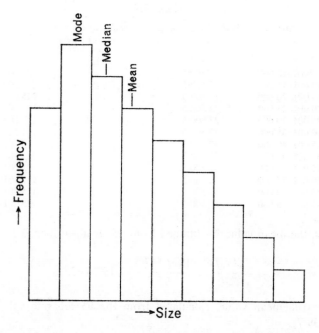

Figure 2.2 Mean, median and modal values in symmetrical and skewed distributions

In such a case, the individual values are put into classes to facilitate calculation of the mean and more particularly the standard deviation, which will be discussed below. There is some loss of accuracy in using grouped data, but this may be a small price to pay for the increased speed of calculation.

With grouped data an error in the calculated mean value arises because the mid-interval value (x_0), or 'half-way point' of each class, is assumed to be the arithmetic mean size of all the items within that class, whereas in most cases, of course, it only approximates to the mean. The overall mean value is calculated from the sum of the products of the mid-interval value (x_0) and the number of units (f) in each class, i.e.

$$\bar{x} = \frac{\Sigma(x_0 f)}{\Sigma f}$$

In the table below, the 95 Municipal Boroughs and Urban Districts in Lancashire have been placed into size classes of 5000 people.

Table 2.1

Class	Mid-interval value (x_0)	Number of units in class (f)	($x_0 f$)
1– 5000	2500·5	12	30 006
5001–10 000	7500·5	17	127 508·5
10 001–15 000	12 500·5	19	237 509·5
15 001–20 000	17 500·5	14	245 007
20 001–25 000	22 500·5	7	157 503·5
25 001–30 000	27 500·5	5	137 502·5
30 001–35 000	32 500·5	5	162 502·5
35 001–40 000	37 500·5	2	75 001
40 001–45 000	42 500·5	5	212 502·5
45 001–50 000	47 500·5	2	95 001
50 001–55 000	52 500·5	3	157 501·5
55 001–60 000	57 500·5	2	115 001
60 001–65 000	62 500·5	2	125 001

Thus, the mean value $\bar{x}$, obtained from the grouped data is

$$\frac{\Sigma(x_0 f)}{\Sigma f} = \frac{1877\ 547 \cdot 5}{95} = 19\ 763 \cdot 7.$$

(The true mean value, calculated from the 95 separate values, is 19 739·7).

With this method of calculation, the value of the product $x_0 f$ may be very large. Where possible, large numbers should be avoided, as they are less convenient to handle than are small ones. Often, this is a very simple matter. For instance, it would be pointless to calculate the mean of 1001, 1002, 1006 and 1009 by adding them and dividing by 4. It would be much simpler to remove the 1000's, calculate the mean of the value by which each exceeds 1000, and then put back the 1000,

$$\text{i.e. } \bar{x} = 1000 + \frac{1 + 2 + 6 + 9}{4} = 1000 + 4 \cdot 5 = 1004 \cdot 5$$

$$\text{rather than } \bar{x} = \frac{1001 + 1002 + 1006 + 1009}{4} = \frac{4018}{4} = 1004 \cdot 5.$$

The large numbers in our problem can be avoided in a similar manner, by assuming one of the mid-interval values to be the mean of the whole set of data. This *assumed mean* or *arbitrary origin* $(\bar{x}_0)$ is subtracted from each mid-interval value (x_0), and the value $(x_0 - \bar{x}_0)$ is divided by the class interval (c) so that the difference (d) for each class is measured in units equal to the class interval, i.e. $d = \dfrac{(x_0 - \bar{x}_0)}{c}$. The number of items in each class (f) is then multiplied by the value of d for that class, and the sum of all these products (Σfd) is calculated. These calculations are set out below:

Table 2.2

Class	Mid-interval value (x_0)	No. of units in class (f)	d $(x_0 - \bar{x}_0)/c$	fd
1– 5000	2500·5	12	−4	−48
5001–10 000	7500·5	17	−3	−51
10 001–15 000	12 500·5	19	−2	−38
15 001–20 000	17 500·5	14	−1	−14
20 001–25 000	22 500·5	7	0	0
25 001–30 000	27 500·5	5	1	5
30 001–35 000	32 500·5	5	2	10
35 001–40 000	37 500·5	2	3	6
40 001–45 000	42 500·5	5	4	20
45 001–50 000	47 500·5	2	5	10
50 001–55 000	52 500·5	3	6	18
55 001–60 000	57 500·5	2	7	14
60 001–65 000	62 500·5	2	8	16
		$n = \Sigma f = 95$		$\Sigma fd = -52$

$$(\bar{x}_0 = 22\ 500 \cdot 5 \qquad c = 5000)$$

The arithmetic mean, calculated from the expression

$$\bar{x} = \bar{x}_0 + c \cdot \frac{\Sigma fd}{n}$$

is

$$\bar{x} = 22\,500 \cdot 5 - \frac{(5000 \times 52)}{95} = 19\,763 \cdot 7.$$

It can be seen that large numbers have been avoided and the correct answer achieved. An arbitrary origin in any of the classes would give the same result, but normally a value of $\bar{x}_0$ which looks as if it is near the overall mean value is chosen.

Here attention must be drawn to the difference between *discrete* and *continuous* variables. Discrete variables can differ only by a constant amount, usually a whole number, whereas continuous variables can take any value within the total range over which they are distributed. Thus, the population of a town can only be a whole number, whereas its area lies on a continuous scale. Similarly a farm cannot possess 41·4 cattle, but it may be 41·4 acres in size or even 41·42 acres, 41·423 acres or any other of an infinite number of possible sizes between 41 and 42 acres. Continuous variables may be converted into discrete ones by taking values to the nearest whole number. Thus, farms of 41·4 acres and 41·6 acres may be assigned the discrete values of 41 and 42 respectively. This conversion technique was applied to the Rousay farms at the time of data collection in 1966.

Continuous and discrete variables which have been grouped into classes must be distinguished carefully, as the class and mid-interval values differ. Continuous class sizes could be 0–99·9̇9̇, 100–199·9̇9̇, etc., with mid-interval values of exactly 50, 150, etc. The equivalent discrete classes (1–100, 101–200, etc.) have mid-interval values of 50·5, 150·5, etc., which are 0·5 higher than those of the continuous classes. The discrete classes of the data above (1–5000, 5001–10 000, etc.) thus have mid-interval values of 2500·5, 7500·5, etc.

Measures of central tendency are a useful method of summarising information, but they are only one method. Two sets of geographical data may be so distributed that their mean values are the same, yet the distributions may differ markedly. One set may cluster closely around the mean whilst the other is spread widely about it (Figure 2.3). Clearly, therefore, in order to summarise information adequately, it is necessary to measure the degree of clustering, or spreading, as well as to determine the measures of central tendency. Merely to say that the mean population of Municipal Boroughs and Urban Districts in Lancashire in 1961 was 19 739·7 does not tell us a great deal about these towns, unless we also have some information about the degree of *variability* or *dispersion*. Obviously, the total *range* of values gives some indication of spread, but two distributions may have the same mean value and the same difference between the largest and smallest values and yet have different 'spreads'.

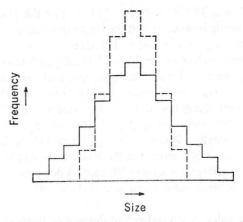

Figure 2.3 Distributions with the same mean value, but different standard deviations

A simple measure of variability is to find the value below which lies a particular percentage of the distribution. Such a value is known as a *percentile*; the nth percentile is the value which divides the total frequency distribution in the ratio $n : 100 - n$. Thus, the median is the 50th percentile. The 25th percentile, the median and the 75th percentile divide the distribution into quarters, and are therefore known as *quartiles*, the difference between the 25th and 75th percentiles being termed the *inter-quartile range*. This is a useful if rather crude measure of the spread, for the smaller the inter-quartile range the more closely does the distribution cluster about the median. If the quartiles fall between observed values, they are given values equal to the arithmetic mean of the numbers on either side. Thus, the lower quartile in a set of 100 observations lies between those values which are 25th and 26th in order of size, and it is given the value $\dfrac{x_{25} + x_{26}}{2}$.

If there are at least 50 values in the whole set, then the inter-quartile range is a fairly accurate measure of their dispersion.

As there were 95 Municipal Boroughs and Urban Districts in Lancashire in 1961, the value of the lower (25th) quartile is equal to the arithmetic mean size of the towns which were 23rd and 24th in order of size, i.e., the lower quartile $= \dfrac{7018 + 7064}{2} = 7041$. The upper (75th) quartile is $\dfrac{26\,726 + 27\,502}{2} = 27\,114$. The inter-quartile range thus is $27\,114 - 7041 = 20\,073$. For the Rousay farms, the lower quartile is 40 acres and the upper quartile is 212 acres; the inter-quartile range is $212 - 40 = 172$ acres.

Another approach to the problem of variability is to consider each unit in turn, compare it with some constant, preferably an 'average', and then summarise the differences from that constant. The difference between the size of any one unit and the 'average' value indicates the *deviation* of that particular unit from this 'average'. Thus the deviation of the smallest Rousay holding (4 acres) from the arithmetic mean size is about 246 acres, and its deviation from the median size is 84 acres. The largest holding (3000 acres) deviates by about 2750 acres from the arithmetic mean size and by 2912 acres from the median.

The mean value of all individual deviations from a given value, which is a measure of the spread of the whole range of sizes, is known as the *mean deviation*. This can be expressed as $\dfrac{1}{n}\Sigma|d|$ where $|d|$ is the deviation of a single unit without regard to its arithmetic sign. For example, $|d| = 3$ denotes that the unit concerned deviates positively or negatively away from the mean or median size by 3 acres, i.e. the holding is 3 acres smaller or larger than the mean or median size. (The mean deviation may be measured from the mean or the median, but is best taken from the median because in that case it is at a minimum.) The mean deviation of the size of Rousay holdings from the arithmetic mean size of 250·04 acres is about 272 acres, and the mean deviation from the median size of 88 acres is about 214 acres.

More important and useful as a measure of spread or dispersion, however, is the mean squared deviation from the mean value, known as the *variance* of the distribution. This is calculated by measuring the deviation of each value from the mean, squaring each deviation, and working out the mean of all these squares. (As, by definition, the sums of the negative and positive deviations from the mean are exactly equal, if they were not squared their sum would be zero.) Thus the variance, σ^2 (sigma squared) may be expressed as

$$\frac{\Sigma(x - \bar{x})^2}{n}.$$

The sum of the squares of the deviations of the 44 individual Rousay farm sizes from the approximate mean value of 250 acres is 15 251 071 and the variance thus is $\frac{15\ 251\ 071}{44}$ or 346 615·26. The square root of the variance is known as the *standard deviation* (σ) of the distribution, sometimes called the *root-mean-square* deviation because of the method of calculation. It indicates the degree of clustering of individual values around the mean, and it is measured in the same units as are the individual items. If the acreage of each farm is used in calculating the standard deviation of Rousay farm sizes, the latter is given in acres. If farm sizes are measured in square miles, then the standard deviation also is in square miles. The standard deviation is expressed by the formula

$$\sigma = \sqrt{\frac{\Sigma(x - \bar{x})^2}{n}},$$

and it can be shown mathematically that an alternative expression is

$$\sigma = \sqrt{\frac{\Sigma x^2}{n} - \bar{x}^2}.$$

This second expression is more convenient as it involves fewer stages in calculation, the individual deviations not being required. For the Rousay farms, the standard deviation is $\sqrt{346\ 615·26}$ or 588·7 acres.

The more closely the individual values in a group cluster about the mean value, the smaller are the variance and the standard deviation. The mean size of the 22 farms on Rousay which cover less than 100 acres each is 35·64 acres. Both the variance (6553·35) and the standard deviation (80·93 acres) of this group of farms, ranging in size from 4 to 76 acres, are less than the values for the remaining 22 farms. These larger farms range in size from 100 to 3000 acres, with a mean value of 464·45 acres, a variance of 459 833 and a standard deviation of 678·1 acres.

We have noted above that the time required for the calculation of the mean from many individual values may be reduced by grouping the data into classes. The calculation of the standard deviation from individual values is even more time consuming than that of the mean, as each value of x has to be squared before it is summed. The grouping of data for the calculation of σ therefore saves even more time. For the Lancashire towns the classes listed previously can be used. The data are set out as before, with one additional column (fd^2).

Table 2.3

Class	Mid-interval value (x_0)	No. of units in class (f)		fd	fd^2
1– 5000	2500·5	12	−4	−48	192
5001–10 000	7500·5	17	−3	−51	153
10 001–15 000	12 500·5	19	−2	−38	76
15 001–20 000	17 500·5	14	−1	−14	14
20 001–25 000	22 500·5	7	0	0	0
25 001–30 000	27 500·5	5	1	5	5
30 001–35 000	32 500·5	5	2	10	20
35 001–40 000	37 500·5	2	3	6	18
40 001–45 000	42 500·5	5	4	20	80
45 001–50 000	47 500·5	2	5	10	50
50 001–55 000	52 500·5	3	6	18	108
55 001–60 000	57 500·5	2	7	14	98
60 001–65 000	62 500·5	2	8	16	128

$$\Sigma fd^2 = 942$$

The standard deviation is calculated from the formula

$$\sigma = c\sqrt{\frac{\Sigma fd^2}{n} - \left(\frac{\Sigma fd}{n}\right)^2}.$$

The standard deviation of the population of Lancashire towns in 1961 thus was

$$5000\sqrt{\frac{942}{95} - \left(\frac{-52}{95}\right)^2}$$

$$= 5000\sqrt{9\cdot9158 - 0\cdot2996}$$
$$= 5000\sqrt{9\cdot6162}$$
$$= 15\ 505 \text{ persons.}$$

The mean values and measures of dispersion for the sizes of holdings on Rousay and towns in Lancashire are summarised in the table below. The extreme values of individual units affect all the measures of dispersion, but the inter-quartile range and standard deviation are not affected to the same degree as is the absolute range.

Table 2.4

	Rousay holdings 1966 (acres)	Lancashire M.B.'s and U.D.'s 1961 (people)	
		Individual form	Grouped form
Arithmetic mean value	250·04	19 739·7	19 763·7
Median value	88	14 474	12 500·5
Absolute range	2996	60 975	60 000
Inter-quartile range	172	20 073	20 000
Mean deviation from arithmetic mean	272	12 128·2	12 322·5
Mean deviation from median	214	11 230·6	11 579·0
Standard deviation	588·7	15 379·4	15 505·0

The mean size of Rousay holdings is 250·04 acres and the standard deviation is 588·7 acres. Although these values together tell us much about the size distribution of the holdings in absolute terms, they cannot be compared directly with some other factor not measured in acres, such as the distribution of herds of cattle. In order that such comparisons can be made, it is necessary to convert the measures in some way. A similar conversion is necessary if we are to compare two different standard deviations. Although we may compare directly the mean number of cattle per holding in Rousay with the mean number of sheep, since one is just a larger number than the other, their standard deviations may not be so compared; a particular standard deviation represents a greater amount of deviation from a small arithmetic mean than from a larger mean. However, the standard deviation may be expressed as a percentage of the mean to give a *coefficient of variation*, which is a relative measure of dispersion, i.e.

$$\text{coefficient of variation} = V = \frac{100\,\sigma}{\bar{x}} \text{ per cent.}$$

The mean number of cattle per holding in Rousay is 37·3 and the standard deviation is 36·0. The mean number of sheep per holding is 53·8 and the standard deviation is 102·5. The coefficients of variation thus are 96·5 and 190·5 for cattle and sheep respectively; since the latter value is the larger, we must conclude that the number of sheep per holding in Rousay is more variable than is the number of cattle.

It is plain that lengthy verbal accounts are not essential for a clear description of the size distribution of units such as the land holdings on Rousay or the towns of Lancashire. A few simple calculations enable us to assess, not only the 'average', 'representative', or 'most characteristic' values of particular data, but also the variability, spread or dispersion of the individual values within the group. With the aid of these numerical expressions, or *parameters*, the characteristics of the size distribution may be described concisely. Determination of these parameters thus may clarify the results of the geographer's investigations without consuming a great deal of his time.

3. The distribution of geographical data

Since this book is concerned with the quantitative approach to the study of geography, it must deal with the measurement of geographic phenomena. The process of measurement inevitably involves errors of various orders of magnitude. Suppose, for example, that we wished to measure from a previously-drawn map the area of a particular farm on Rousay. One possible method would be to lay a tracing of the farm outline on a piece of squared paper and to count the enclosed squares. If many students of equal ability were to measure the

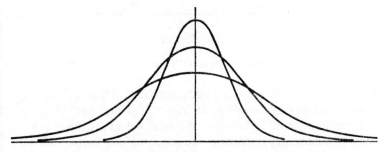

Figure 3.1 Three normal curves

farm size using this method, they would arrive at different values for the area because areas of parts of squares would have to be estimated—the farm boundary would cut across some squares. Some students would underestimate the area and others would overestimate it. Errors involved in measurement would have to be assessed so that we could decide which figure should be accepted as representing the area of the farm.

It appears that errors have a particular distribution, being spread equally about the true value—the mean of the measurements is the most probable true value, and most values cluster about the mean. This *normal law of errors* describes a symmetrical, bell-shaped distribution curve (Figure 3.1). There is a greater probability that any single measurement will lie within a range of values close to the mean

19

than within an equal range further from it, very inaccurate estimates being more improbable than more accurate ones. Because of this, the 'normal' curve sometimes is called the *normal probability curve*.

The term 'normal' has been used above in a special way, different from everyday use. The application of the description 'normal' to a distribution does not imply that other distributions are in any way abnormal. There is no one normal distribution, but a family of curves, all of which possess the bell-shaped normal form (Figure 3.1). Statistical normality is a property of a distribution, not of individual

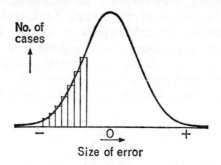

Figure 3.2 Geometrical approximation of a curve to a histogram.

values within it: one value may not be described as 'normal', although a set of values may be said to be 'normally distributed'.

If the number of occurrences of certain values are plotted in the form of a frequency distribution diagram, then the frequencies must be discrete, for a particular value can occur only a whole number of times. A frequency distribution of cattle on Rousay, for instance, will indicate the number of farms on which any one number of cattle is kept; the 'frequency' of 27 cattle is three. Similarly, the frequency distribution of errors in measuring a single area will show the number of times on which a given size of error occurs. A frequency diagram thus is a histogram. However, if the bars are made extremely narrow they merge and the plot becomes indistinguishable from a curve (Figure 3.2) which is much easier to handle mathematically. This *geometrical approximation* is justified in that the area under the curve approximates closely to the sum of the areas of the individual bars in the histogram. In theory, the normal curve extends infinitely in both directions, but most (99·99 per cent) of the area below it is bounded by four standard deviations on either side of the mean. Similarly, 99·73 per cent of the total area is within three standard deviations, 95·45 per cent within two, and 68·27 per cent within one (Figure 3.3).

In addition to error distributions, many other distributions conform closely to the normal curve, and therefore can be treated as having normal characteristics; this makes the normal curve a very useful tool in statistical analysis. Its use is not restricted to data which are normally distributed; many sets of data may approximate to normality after *transformation*, a property which justifies the work involved. A very common way of transforming data is to use the

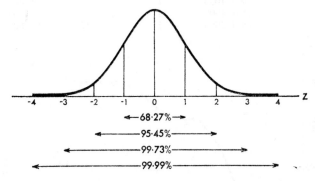

Figure 3.3 The normal curve, showing areas enclosed by standard deviations

logarithms of the sizes of the phenomena under consideration in place of the sizes themselves; the frequencies remain the same. The normal curve is also of fundamental importance in sampling, which is discussed in Chapter 4.

The assessment of the normality of a distribution can be achieved rapidly with the aid of a special type of graph paper, known as *probability paper*, but first the distribution must be expressed as a *cumulative frequency percentage curve*. A cumulative frequency curve shows the number of items or values above or below a particular level. It is constructed from a simple histogram by adding the number in each class to those in the classes above or below. It can be made into a percentage curve by substituting for the absolute numbers the percentage of the total which each class contains. The cumulative frequency curve for Rousay farm sizes may be determined by calculating the percentage of all farms below certain sizes, say every 50 acres, and then adding successive percentages. (Shown on the next page.)

Table 3.1

Size (acres)	No. of farms	Percentage of total number	Cumulative percentage
1–50	15	34·10	34·10
51–100	8	18·18	52·28
101–150	5	11·37	63·65
151–200	5	11·37	75·02
201–250	3	6·82	81·84
251–300	0	0·00	81·84
301–350	1	2·27	84·11
351–400	2	4·54	88·65
401–450	1	2·27	90·92
551–600	1	2·27	93·19
951–1000	1	2·27	95·46
1851–1900	1	2·27	97·73
2951–3000	1	2·27	100·00

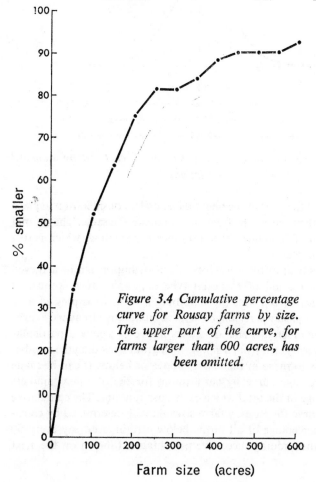

Figure 3.4 Cumulative percentage curve for Rousay farms by size. The upper part of the curve, for farms larger than 600 acres, has been omitted.

Farm size   (acres)

The lower part of the cumulative percentage curve for Rousay farms by size is plotted in Figure 3.4. The horizontal scale would have to be extended to 3000 acres in order that the entire distribution

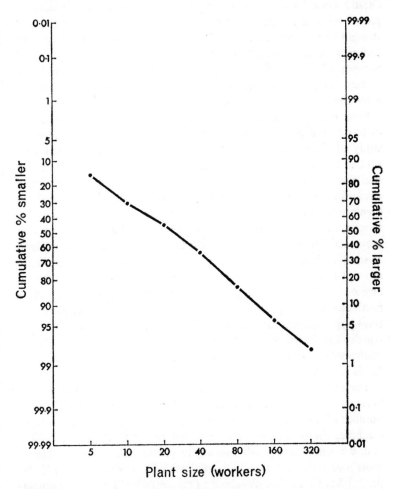

Figure 3.5 The log-normal size distribution of a sample of Manchester clothing firms

(100 per cent) could be shown on the arithmetic scale used. On probability paper the vertical (i.e. frequency) scale has been adjusted so that any normal distribution will appear as a straight line if it is in cumulative percentage form. The horizontal axis can be divided

arithmetically or logarithmically depending on whether or not the data need be transformed to achieve a straight line. Figure 3.5 shows the distribution by size of some of the clothing firms in the Manchester conurbation, drawn originally on logarithmic probability paper; with this transformation the distribution approximates to a straight line. In this case firms *of* or *below* a given size all have been summed, or accumulated, and expressed as a percentage of the total number of firms: the distribution thus is shown in a cumulative percentage form. Because of the logarithmic transformation, the firms can be treated as *log-normally distributed*.

Since the geographer frequently is unable to draw firm conclusions or to quote definite, or even approximately accurate, results from a study of the data available to him, he has to be concerned with probabilities. He must attempt to interpret the data in such a way that the *probable* outcome or result of a particular course or set of circumstances becomes evident. Although much geographical information cannot be measured quantitatively, it is frequently possible to measure the probability that a particular conclusion based upon it is correct. The number of farms, cattle and arable acres on an island as small as Rousay can be counted, but it would be impracticable to attempt to count all the farms, cattle and arable acres in Britain at any one time. However, we can determine the probability that a particular selected group, or *sample*, of farms is representative of the total population of farms, and so we can assess the reliability of any conclusions based on this sample. In other words, we can state the *statistical significance* of a conclusion, the probability that the hypothesis is correct.

The probability that an event will occur is expressed as the ratio between the total number of ways in which it can occur and the total number of *all* possible events in the given circumstances. Thus, for example, the probability that a tossed coin will display a head is $\frac{1}{2}$, or 0·5, since there is only one way in which a head can occur uppermost and only two possible outcomes of the event. Similarly the probability that a dice will show a four (or indeed any other number) on the uppermost face is $\frac{1}{6}$ or about 0·17. If we were to select one farm from those on Rousay, there is a probability of $\frac{15}{44}$ or about 0·34 that it would be not more than 50 acres in size, since 15 of the 44 farms on the island are not larger than 50 acres. It is important to realise that, although the Rousay farms may be divided into two categories, 'larger than 50 acres' and 'not larger than 50 acres', the chance of a single selection falling into one particular category is not the same as the chance that it will fall into the other, as there are not equal numbers of farms in each. Although there is a 1 in 2 chance that a tossed coin will display a head and a 1 in 2 chance that it will display a tail, the chance that a selected Rousay farm will not be larger than 50 acres is only 1 in about 2·9 (15 in 44), whilst there is a one in about 1·5 chance (29 in 44) that it will be larger than 50 acres. With the coin, then, the possibilities are equally probable or *equiprobable*; with the farm sizes they are not.

The Rousay farms, which may be termed a *set* (a well-defined collection of objects), can be divided into *subsets*, each composed of a number of *elements* (farms) which have a characteristic or characteristics in common. The subsets might be chosen in accordance with a wish to classify the farms. For instance, we might classify as 'large' those farms bigger than 50 acres, those with more than 50 cattle or 50 sheep, or those where any combination of these characteristics holds. In Figure 3.6a, each element of the set of 44 Rousay holdings in 1966 is represented by a single square. There is a subset of 29 farms which have in common the fact that they are larger than 50 acres (Figure 3.6b), a subset of 12 farms with more than 50 cattle (Figure 3.6c), a subset of 15 with more than 50 sheep (Figure 3.6d) and a subset of 14 which are not larger than 50 acres and do not possess more than 50 cattle or 50 sheep (Figure 3.6e). By superimposing each of these groups on the outline of the 44 holdings, a diagrammatic representation of the farms may be produced, in which the characteristics of each farm with regard to all three criteria of 'largeness' are shown (Figure 3.6f).

This diagram may be used to determine the probability that a single farm selected at random will possess any one or any combination of the three independent characteristics illustrated. Thus, the probability that a farm will not have more than 50 sheep, but will be larger than 50 acres and will have more than 50 cattle is $\frac{2}{44}$, or about 0·05. Similarly, the probability that a farm will be larger than 50 acres and will have more than 50 sheep is $\frac{4}{44} + \frac{10}{44} = \frac{14}{44}$, or 0·32. The probability that a single Rousay farm selected at random from the list possesses at least one of the characteristics which make it 'large' is $\frac{30}{44}$, or 0·68.

The subsets represented in Figures 3.6b, c and d may be redrawn as circles which overlap in a manner similar to the blocks in Figure 3.6f; the form of presentation is termed a *Venn diagram* (Figure

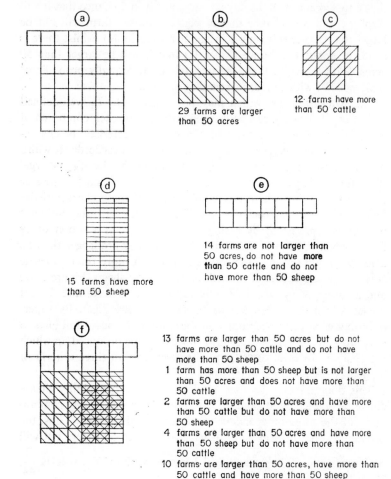

Figure 3.6 The attributes of Rousay farms

3.7). Once again, the arrangement of the data may be used to assess probabilities. For instance, four farms (18, 25, 33, 34) lie within the intersection of the subsets A ('larger than 50 acres') and C ('with more than 50 sheep'), but outside the subset B ('with more than 50 cattle'); the probability that a farm selected at random from the list

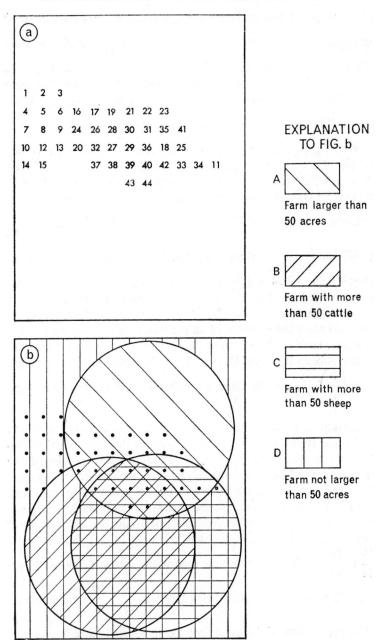

EXPLANATION
TO FIG. b

A

Farm larger than
50 acres

B

Farm with more
than 50 cattle

C

Farm with more
than 50 sheep

D

Farm not larger
than 50 acres

Figure 3.7 Venn diagram of Rousay farms

of Appendix A will be larger than 50 acres, possess more than 50 sheep, but not more than 50 cattle, therefore is $\frac{4}{44}$ or 0·09. The Venn diagram also permits calculation of the probability that a selected unit possessing one particular characteristic will also possess another one. For instance, given that a chosen farm belongs to subset A, the probability that it also belongs to subset B is $\frac{12}{29}$.

Where categories are mutually exclusive, their individual probabilities of occurrence or selection may be added (*the law of addition*). Thus, no farm can fall within both subset A ('larger than 50 acres') and subset D ('not larger than 50 acres'). In selecting a single farm at random, the probabilities of its falling into these subsets are $\frac{29}{44}$ (0·66) and $\frac{15}{44}$ (0·34) respectively. The total probability of outcomes thus is 0·66 + 0·34 = 1. Similarly, a tossed coin is *certain* to display either a head (probability $\frac{1}{2}$) or a tail (probability $\frac{1}{2}$); once again, therefore, the total probability of outcomes is 1. In fact, this is the case whenever the outcome of an event is certain. If an event is impossible, such as the selection of a single Rousay farm which lies within subsets B, C and D, then the probability of its occurrence is zero $\left(\frac{0}{44}\right)$. Clearly, therefore, all probability values must lie between 0 and 1.

If events are independent of each other, the probability that both will occur is not the sum but the product of their individual probabilities (*the law of multiplication*). For example, having selected one farm from the Rousay list, we may go on to make a second selection from the total of 44. Each farm in the list could be 'paired' with itself or with any one other farm, since any combination of two may arise from the two independent selections. Thus the probability of both farms being not larger than 50 acres in size (subset D) is $\frac{15}{44} \times \frac{15}{44}$, or about 0·11 and the probability of one being not larger than 50 acres and the other being larger is $\frac{15}{44} \times \frac{29}{44}$, or about 0·22.

(The probability of the same farm being selected twice is $\frac{1}{44} \times \frac{1}{44}$, or about 0·00052). For similar reasons, the probability of tossing two heads in two successive attempts is $\frac{1}{2} \times \frac{1}{2}$, or 0·25; the two events are completely independent of each other as the outcome of the second is not affected by the outcome of the first.

We have seen already that events, characteristics or subsets may be neither mutually exclusive nor independent, but interlocked, and that we can calculate the probability of their occurring together. It is also possible to determine the probability that at least one of two characteristics will be possessed by a selected farm, for this is equal to the sum of the probabilities that each will be possessed minus the probability that both will be possessed, e.g. the probability that a selected farm will be larger than 50 acres (subset A), have more than 50 sheep (subset C), or both (subsets A and C) is $\frac{29}{44} + \frac{15}{44} - \frac{14}{44} = \frac{30}{44}$.

The probability that a selected farm lies within both subset A and subset B is $\frac{12}{44}$. This is the same as $\frac{29}{44} \times \frac{12}{29}$, the probability that the farm lies within subset A multiplied by the probability that it lies within subset B if it does lie in subset A; this illustrates the law of multiplication for independent events already mentioned.

The theory of probability is of fundamental importance in statistical analysis where complete information about a particular situation is not available or cannot be collected practicably and where, as a result, sampling is necessary; it should be stressed that the availability of complete information is the exception rather than the rule. To be able to assess the probability of selecting units with certain characteristics is of the utmost importance when taking and interpreting samples, and the implications of probability theory for sampling method are discussed in Chapter 4.

If the area beneath a normal curve is taken as unity, it may be regarded as total probability. The curve then is useful in calculating the probability that any individual item in a set will be above or below a given value, although it cannot be used to calculate what proportion of several selected items will be above or below that value (Appendix D). One of the most valuable uses of the normal curve is in climatic studies. If climatic data for a particular station, such as rainfall amounts, can be shown to be normally distributed, then the probability of receiving more or less than a particular amount can be calculated. From Figure 3.5 we can deduce that 30 per cent of the group of Manchester clothing firms employ fewer than

10 workers. The probability that any individual firm selected at random from the group will employ fewer than 10 workers therefore is 0·30 and the probability that the firm will employ not fewer than 10 is 0·70. It does not follow that, if we were to select 10 firms, three of them would employ fewer than 10 workers and seven not fewer than 10, although with a larger selection, say 100 firms, the proportions 3 : 7 would be more likely to occur.

With a selection of n firms, it is possible to calculate the probability that any number of them will have fewer than 10 employees. The *binomial frequency distribution* expresses the relative frequency of two mutually exclusive conditions—in this case, 'fewer than 10 employees' and 'not fewer than 10 employees'. The probabilities of selecting a firm with fewer and with not fewer than 10 employees may be denoted by p and q respectively. As p and q are mutually exclusive, it follows from the law of addition that $(p + q) = 1$. When $(p + q) = 1$, then $(p + q)^n = 1$. The expansion of $(p + q)^n$, where n is the number of firms selected, describes the binomial distribution.

Expanding $(p + q)^n$ by simple algebra gives:

$$(p + q)^1 = p + q$$
$$(p + q)^2 = p^2 + 2pq + q^2$$
$$(p + q)^3 = p^3 + 3p^2q + 3pq^2 + q^3$$

It can be seen that, when $(p + q)^n$ is expanded, the powers of p decrease in steps of 1 from n to 0 from left to right, and the powers of q decrease in a similar manner from right to left. A general form for the expansion thus is:

$$(p + q)^n = p^n + C_1 p^{n-1}q + C_2 p^{n-2}q^2 + \ldots + q^n$$

where C_1, C_2, etc., are the *coefficients* of the expansion. The values of the coefficients are indicated by a table known as *Pascal's Triangle*. In this, successively lower lines of numbers are calculated by the addition of adjacent pairs of numbers in the line immediately above, the top line being composed of two values of 1 bordered on each side by an infinite number of zeros (not shown in the diagram below).

n				Values of coefficients				
1				1		1		
2				1	2	1		
3			1	3	3	1		
4		1	4	6	4	1		
5	1	5	10	10	5	1		
6	1	6	15	20	15	6	1	

If we were to select 5 firms at random from the group of Manchester clothing firms, the total of the probabilities of selecting various combinations would be $(p + q)^5 = 1$, or

$$p^5 + 5p^4q + 10p^3q^2 + 10p^2q^3 + 5pq^4 + q^5 = 1.$$

The probability of selecting a firm with fewer than 10 employees (p) is 0·30, and the probability of selecting one with not fewer than 10 employees (q) is 0·70. Thus, the probability that all 5 selected firms will have fewer than 10 employees is $p^5 = (0·30)^5 = 0·002$ and the probability that all 5 will employ not fewer than 10 workers is $q^5 = (0·70)^5 = 0·168$. This is another simple application of the law of multiplication for independent events. (In fact, the probabilities will be slightly altered by the non-replacement of firms once drawn, i.e. if the first drawn employs less than 10, the number of firms in this category available for the second draw is reduced by one. However, the alteration of the 'odds' is so small that the law remains almost completely valid.)

The terms of the binomial expansion represent all the possible combinations of p and q, from 5 occurrences of p at one end to 5 of q at the other. The probability of selecting 4 firms with fewer than 10 workers and one with not fewer than 10 is

$$5p^4q = 5.(0·30)^4.(0·70) = 0·028.$$

Similarly, the probability of selecting *at least* three firms with fewer than 10 workers, a condition satisfied by the first three terms in the equation is:

$$p^5+5p^4q+10p^3q^2 = (0·30)^5+5.(0·30)^4.(0·70)+10.(0·30)^3.(0·70)^2$$
$$= 0·163.$$

Thus, the binomial distribution is used in estimating what proportion of a set of data is larger or smaller than a critical value or possesses a particular qualitative characteristic, and so, like the normal distribution, it is of great use in sampling.

4. Estimating information

Because of the large number of firms involved, a geographical study of the Manchester clothing industry could not involve a detailed survey of every unit. Although it would be preferable to have information about every firm concerning the total number of employees, the proportion of female workers, the quantity of goods of each type produced and many other characteristics, it is unlikely that such information could be obtained, even with the aid of questionnaire surveys and the expenditure of much time and effort in the field. In order to make general statements about Manchester's clothing industry, therefore, we would have to arrive at our conclusions by studying a representative selection of firms. As time and expense often prevent the geographer studying, testing or obtaining complete information about all the people, places or objects in which he is interested, he must rely in very many studies upon examination of such selected representatives.

Although different clothing firms in the Manchester area, like individual elements within any set, are likely to display different characteristics, different reactions to given circumstances and so on, the group as a whole may conform in some ways to a particular pattern, so that it may be possible to estimate the characteristics or reactions of the group (*parent population*) from studies of a representative section (*sample*). If a definite pattern can be recognised within a sample of, say, clothing firms, and can be measured and related to a theoretical distribution pattern, such as the normal distribution, then the information may be applied to the parent population by the use of probability theory.

It is the correct choice of a sample from the population which enables the geographer to obtain information which otherwise would be beyond his reach. Unfortunately, conclusions drawn from studies of samples not representative of the total population are far from uncommon in geographical literature. If no attempt is made to check the assumption that the sample is typical of the total population, the conclusions drawn about the latter may be grossly in error. Clearly, it would be unwise to select for study the 100 firms which were first in alphabetical order and to treat them as 'characteristic' of all clothing firms in Manchester, for this group might, by chance, be biased in favour of certain subsets, such as large firms or firms manufacturing men's outer wear, so that the importance of large firms or men's outer wear manufacturers within the total population would be exaggerated. Similarly, to treat the 22 holdings which lie wholly within the north-eastern half of the island as representative of the Rousay farms as a whole would lead to false conclusions. For instance, both the mean size of this sample (142·32 acres) and the variability of size of the farms ($\sigma = 130·7$ acres) are considerably less than the mean size of all Rousay farms (250·04 acres) and the overall variability of farm sizes ($\sigma = 588·7$ acres).

Much thought must go into the task of selecting an adequate sample. In deciding upon the optimum size to be used for a particular investigation, the size and nature of the population from which the sample has to be drawn have to be considered. The nature of the questions to which answers are sought is another important consideration. In a survey of differences between firms engaged in the manufacture of men's outer wear, for instance, there would be little point in drawing a sample from a population consisting of all clothing firms—the population sampled should consist of men's outer wear manufacturers only. If it is desired to compare per capita output from men's outer wear firms employing six or fewer workers with that from other firms, one sample will have to be drawn from the population of those men's outer wear manufacturers who employ no more than six people.

For two different samples drawn from the same population, the probability that the larger will be within a certain margin of a particular population parameter is greater, because items causing variability ('the exceptions to the rule') become relatively less important or less frequent with larger samples. Nevertheless, although a larger sample is likely to be more representative than a smaller one, because it approaches more closely the size of the population itself, investigations of very large samples necessitate the expenditure of much time and labour, and a smaller sample therefore may be more desirable. Although there is 'safety in numbers', the degree to which a sample represents accurately a given population parameter is a much more important consideration than is the actual size of the sample. In other words, sample size is not governed primarily by the size of the population but by the required degree of accuracy.

The most efficient samples are those from which the population parameter of interest may be estimated *within a specified degree of possible error* (the sample estimate of a parameter generally is known as a *statistic*); the acceptable degree of error, e.g. ± 5 per cent, will depend upon the particular data being studied. As has been noted above with regard to the Rousay farms, different samples drawn from the same population are likely to have different mean values, which themselves differ somewhat from the population mean. If many samples were taken, a frequency distribution diagram of their means could be drawn, and the sample means would have their own mean, standard deviation, etc. If all the possible samples of a given size, n, were drawn from a particular population, their mean value would equal the mean of the population itself. If $n = 1$, of course, the distribution of the sample means is identical with the distribution of the individuals which make up the parent population.

The distribution of a particular statistic (mean, standard deviation, etc.) of all possible samples of a given size is known as the *sampling distribution* of that statistic, and the standard deviation of this distribution is known as the *standard error* of the statistic. Thus, the distribution of the mean sizes of all possible samples of 100 Manchester clothing firms would be called the sampling distribution of the mean size of firms in the Manchester clothing industry, and the standard deviation of the distribution of the sample means would be termed the standard error of the mean. If all possible samples of size n are drawn from a population (size N), then the mean and standard deviation of the sampling distribution of the means may be related to the mean and standard deviation of the population by the expressions:

$$\bar{x}_{\bar{x}} = \bar{X}$$

and

$$\sigma_{\bar{x}} = \frac{\sigma}{\sqrt{n}}$$

where $\bar{x}_{\bar{x}}$ and $\sigma_{\bar{x}}$, respectively, are the mean and standard deviation of the sampling distribution of the means, and $\bar{X}$ and σ are the population mean and standard deviation.

We have seen (Figure 3.3) that 99·73 per cent of the total area under the normal curve lies within three standard deviations of the mean, 95·45 per cent within two standard deviations, and 68·27 per cent within one. We have also noted that the curve may be used to calculate the probability that a particular element in a set will be above or below a given value, by assuming the area beneath the curve to represent total probability (Figure 3.3). If a sample is large ($n \geqslant 30$), the sampling distribution of any statistic such as the mean, standard deviation, etc., is approximately normal. Even when the population is not normally distributed, the sampling distribution will approximate to normality provided the sample is large and the population at least twice the size of the sample. This is a useful property since, from the known properties of the normal distribution, it becomes possible to estimate the probability that a particular statistic will be within a given range. For example, we can be confident of finding the mean size of a particular sample of Manchester clothing firms within one standard error about 68·27 per cent of the time and within three standard errors about 99·73 per cent of the time. Thus, we may call these intervals the 68·27 per cent and 99·73 per cent *confidence intervals* for estimating the mean of the sampling distribution of Manchester clothing firms. The end values of these intervals ($\bar{x} \pm \sigma_{\bar{x}}$, $\bar{x} \pm 3\sigma_{\bar{x}}$) are called the 68·27 per cent and 99·73 per cent *confidence limits*. Similarly $\bar{x} \pm 1{\cdot}96\sigma_{\bar{x}}$, $\bar{x} \pm 2{\cdot}58\sigma_{\bar{x}}$ and $\bar{x} \pm 0{\cdot}67\sigma_{\bar{x}}$ are 95 per cent (0·95), 99 per cent (0·99) and 50 per cent (0·50) confidence limits for $\bar{x}$. The percentage confidence may be termed the *confidence level* and the numbers in the confidence limits (1·96, 2·58, etc.) may be called *confidence coefficients* or *critical values* (z_c). The quantity $0{\cdot}67\sigma_{\bar{x}}$ is known as the *probable error* of the estimate of the mean of the sampling distribution. Clearly, confidence limits for the estimation of *any* population parameter obey these rules, and are given by $A \pm z_c\sigma_A$, where A denotes the parameter concerned. Thus, the confidence limits for the population mean are

$$\bar{x} \pm z_c \frac{\sigma}{\sqrt{n}}$$

given that the sample is large ($n \geqslant 30$). Strictly speaking, this expression is valid only when sampling is from a population of infinite size, or when sampling is with replacement, i.e. a unit, once selected, is replaced and may therefore be selected again. In geographical studies, sampling usually is without replacement; a firm once selected from the Manchester list cannot be selected again—the second firm is drawn from 825 rather than the original 826, and the 100th firm is drawn from 726. When sampling is without replacement from a finite population, it becomes necessary to modify the expression for the confidence limits, which becomes

$$\bar{x} \pm z_c \frac{\sigma}{\sqrt{n}} \sqrt{\frac{N-n}{N-1}}$$

Clearly, so long as the population (N) is several times larger than the sample (n), the correction is not a major one. In many cases, it may be ignored for practical purposes.

In order to estimate, with particular confidence limits, the mean size and standard deviation of the population of Manchester clothing firms from the characteristics of a sample, using the above expression, it would be necessary to know the population mean and standard deviation. These, of course, are unknown. The difficulty can be overcome if the mean and standard deviation of the sample are used, with a suitable correction, instead of those of the population. The correction factor, k, known as *Bessel's correction*, has the value $\sqrt{\dfrac{n}{n-1}}$ and converts the sample standard deviation (s) into the *best estimate* ($\hat{\sigma}$) of the population standard deviation (σ), that is $\hat{\sigma} = s \times k$. As the confidence limits when sampling with replacement are given by $\bar{x} \pm z_c \dfrac{\sigma}{\sqrt{n}}$, they may be expressed as $\bar{x} \pm z_c \dfrac{\hat{\sigma}}{\sqrt{n}}$, or $\bar{x} \pm z_c \dfrac{s.k}{\sqrt{n}}$. At the 95 per cent level, $z_c = 1.96$ and the confidence limits are given by

$$\bar{x} \pm 1.96 \frac{s.k}{\sqrt{n}} \text{ or } \bar{x} \pm 1.96 \frac{s}{\sqrt{n}} \sqrt{\frac{n}{n-1}} = \bar{x} \pm \frac{1.96s}{\sqrt{n-1}}.$$

When sampling without replacement, this expression becomes

$$\bar{x} \pm \frac{1.96s}{\sqrt{n-1}} \sqrt{\frac{N-n}{N-1}}.$$

There are 826 (N) firms in the Manchester clothing industry. From a sample of 50 (n), therefore, the estimated mean size at the 95 per cent level is $\bar{x} \pm \dfrac{1.96s}{\sqrt{49}} \sqrt{\dfrac{776}{825}}$, or $\bar{x} \pm 0.27s$. As pointed out above, the value of the correction factor $\sqrt{\dfrac{N-n}{N-1}}$ is so close to unity (0.97) that, in practice, it could be ignored.

Although the sampling distributions of many statistics are approximately normal if samples are large ($n \geqslant 30$), this approximation does not hold for small samples and becomes worse as n decreases. The difference between the mean of a random sample ($\bar{x}$) and the mean of the population from which it is drawn ($\bar{X}$) may be 'standardised' by expressing it in terms of the standard deviation of the population (σ). This is done by calculating the statistic $z = \dfrac{\sqrt{n}\,|\bar{x} - X|}{\sigma}$.

However, when working with samples, σ is unlikely to be known, so that the standard deviation of the sample (s) again must be used as an estimate of σ. In this case the 'standardised difference' is $t = \dfrac{\sqrt{n}|\bar{x} - \bar{X}|}{s}$. If t is calculated for all samples of size n drawn from a normal, or approximately normal, population, the sampling distribution of t may be determined. This distribution, known as *Student's t distribution*, (Appendix F), closely approximates a normal distribution when n is large, but not when n is small. Sampling distributions of statistics for small samples are Student's t distributions.

Efficiency in sampling lies in choosing a sample of the minimum size which permits estimation of a particular parameter at a given confidence level within a chosen range. For example, the mean size of the random sample of Manchester clothing firms listed in Table 4.1 is $55 \cdot 64 \pm 22 \cdot 90$ employees per firm at the 95 per cent level; the standard deviation (s) is $84 \cdot 32$ and the standard error ($\sigma_{\bar{x}}$) is $11 \cdot 92$. For the range of the estimate to be reduced whilst the same confidence level is maintained, the size of the sample must be increased. If the acceptable range is 15 on each side of the mean, the minimum size of the sample may be calculated as follows:

$$z_c \frac{\sigma}{\sqrt{n}}\sqrt{\frac{N - n}{N - 1}} = z_c \frac{s}{\sqrt{n - 1}}\sqrt{\frac{N - n}{N - 1}} = 15.$$

Therefore, as $z_c = 1 \cdot 96$ and $s = 84 \cdot 32$,

$$\frac{(1 \cdot 96)(84 \cdot 32) \times \sqrt{826 - n}}{\sqrt{n - 1} \qquad \sqrt{825}} = 15$$

and $\qquad (1 \cdot 96)^2 (84 \cdot 32)^2 (826 - n) = (15)^2 (825)(n - 1)$

so that $\qquad (3 \cdot 84)(7109 \cdot 86)(826 - n) = (225)(825)(n - 1)$

and $\qquad 2\,256\,073 \cdot 74 - 27\,313 \cdot 24n = 185\,625n - 185\,625$

and $\qquad 22\,560\,734 \cdot 74 + 185\,625 = (18\,625 + 27\,313 \cdot 24)n$

so that $\qquad 22\,746\,359 \cdot 74 = 212\,938 \cdot 24n$

or $\qquad n = 106 \cdot 82.$

Thus, 107 is the minimum sample required to establish the mean size of firms at the 95 per cent level to within 15 on either side of the mean. Clearly, if the size of firms was less variable (i.e. if s had a lower value), the range about the mean would be lower than this at the 95 per cent level.

Table 4.1

Random sample of 50 Manchester clothing firms

Size of firm (no. of employees)	2	3	4	5	6	7	9	10	12	13	14	15	16
No. of firms	1	4	1	1	2	1	1	2	1	1	1	2	1
Size of firm	20	21	22	23	24	25	30	32	35	40	45	50	58
No. of firms	1	1	1	1	1	1	2	2	2	2	1	1	1
Size of firm	60	66	70	88	90	100	120	150	160	190	200	510	
No. of firms	2	1	1	1	1	1	1	1	1	1	2	1	

In drawing a sample of 107 firms from the total population of 826, the proportion sampled, or *sampling fraction* (f) is $\frac{107}{826}$ or $1:7.72$ (0·130). There are many ways of selecting 107 firms. As we have a full list, each can be allocated a number and the selection can be made by using a *random numbers table* (Appendix E). Because there are 826 firms, three digits have to be used, giving numbers 1 to 1000 (001 being 1 and 000 being 1000). All numbers greater than 826 are discarded. The 107 firms whose list numbers correspond to the random numbers drawn are a completely random and representative sample, as no consideration other than chance can have affected the selection. A random sample of 107 firms indicated a mean size of 47·61 ± 12·29 employees at the 95 per cent level (Table 4.2).

Table 4.2

Category	Number of firms	Mean size	Standard deviation
Rainwear	36	61·47	80·83
Men's outer wear	10	7·90	10·16
Ladies' outer wear	14	48·64	70·06
Shirts	16	30·69	40·98
Dresses	31	52·58	77·67
TOTAL	107	47·61	69·54

Table 4.3

Category	Number of firms	Mean size	Standard deviation
Rainwear	37	57·27	55·24
Men's outer wear	7	27·71	34·96
Ladies' outer wear	19	24·21	26·13
Shirts	18	58·94	93·96
Dresses	37	40·46	69·15
TOTAL	118	45·17	63·51

An alternative to random sampling is to take away every seventh firm on the list, to give a *systematic sample* of 118 firms. (To take every eighth firm would give a sample of only 103 firms, less than the required minimum number.) Once collected, a systematic sample is treated in the same way as a random one. A systematic sample of 118 firms drawn from the list of Manchester clothing firms indicated a mean size of $45\cdot17 \pm 10\cdot66$ employees at the 95 per cent level (Table 4.3). In selecting a systematic sample, care must be taken to ensure that the sample interval does not coincide with some regular fluctuation of the whole data at the same interval. This is extremely unlikely to occur with an alphabetically ordered list, but climatic data, or other statistics which vary regularly with time, may present difficulties to systematic sampling.

The Manchester clothing firms can be classified in five subsets, or *strata*, representing different types of manufacture. In order to be truly representative, a sample should include some firms from each category. By taking a random sample from each category, a *stratified sample* is obtained. If we knew the total number of firms in each category, we could ensure that the sample included the correct proportions by using a constant sampling fraction. For instance, the samples represented in Tables 4.2 and 4.3 suggest that the approximate proportions of firms of different types within the Manchester clothing industry are: Rainwear 32 per cent, Men's outer wear 8 per cent, Ladies' outer wear 15 per cent, Shirts 15 per cent and Dresses 30 per cent. Suppose, then, that we knew the actual numbers of firms in each category to be as shown in Table 4.4 and that we wished to draw a stratified sample of 107 firms, we would determine how many firms from each subset should be included in our sample by keeping constant the value of the sampling fraction, f, at $\frac{1}{7.72}$.

Table 4.4

Category	No. of firms in population	No. to be included in sample
Rainwear	264	34
Men's outer wear	66	9
Ladies' outer wear	124	16
Shirts	124	16
Dresses	248	32
TOTAL	826	107

In fact, the value of f has to be altered slightly for some categories, since a whole number of firms must be included in the sample. The error so introduced, however, is likely to be very small. If we had a list of firms broken down into categories, we could select at random

34 from the 264 making rainwear, 9 of the 66 making men's wear, etc. The overall mean size of Manchester clothing firms could be estimated by treating the stratified sample as a random sample of 107 firms.

Because the samples from different strata are of different sizes, the standard error of the mean varies, so that the accuracy of the estimates of the mean firm size in each stratum is not constant. Clearly, as there are five categories and only 107 firms (Table 4.2) or 118 firms (Table 4.3), some of the subsets must be small (< 30). For these, therefore, the distribution cannot be regarded as normal, and must be treated as a Student's t distribution. In this case, confidence limits for population means are given by

$$\bar{x} \pm t_c \frac{s}{\sqrt{n-1}}$$

where the values $\pm t_c$ are *critical values* or *confidence coefficients*. At the 95 per cent level, values of t_c for samples of different sizes are:

Sample size	2	5	10	15	20	25
t_c at 95 per cent level	12·71	2·78	2·26	2·14	2·09	2·06

The confidence limits at the 95 per cent level for the mean sizes oj firms represented in Tables 4.2 and 4.3 are given in Table 4.5.

Table 4.5

	Random sample	Systematic sample
Rainwear	61·47 ± 26·20	57·27 ± 17·65
Men's outer wear	7·90 ± 7·65	27·71 ± 33·54
Ladies' outer wear	48·64 ± 41·97	24·21 ± 12·93
Shirts	30·69 ± 22·54	58·94 ± 48·08
Dresses	52·58 ± 27·28	40·46 ± 22·09

The confidence limits generally are too wide for the actual mean values to be significant. In part, this reflects the small number of firms within some of the subsets and the consequent 'weighting' of the mean by individual firms. Thus, the largest of the 10 men's outer wear firms in the random sample employed 35 workers, whilst one of the 7 such firms in the systematic sample employed 100. Similarly, 420 of the 681 employees in ladies' outer wear manufacture selected by the random sampling are employed by two of the fourteen firms.

If a constant acceptable range for the estimates of stratum means is desired, a *variable sampling fraction* must be used; the sampling fractions for some strata must be increased so as to increase the number of firms in those strata. This is particularly important where f gives a very small sample, as in the men's outer wear stratum of Table 4.4. To increase the sampling fraction to $\frac{1}{2}$ in this case would involve collecting information about 33 firms rather than 9. The improvement in the results of the survey would more than balance the extra work involved. In an extreme case, every unit within a stratum of very small size would have to be considered. If, as in fact is the case with regard to the Manchester firms, we did not know the proportions of the total population within various strata, it might be necessary to visit very many firms in order to secure a modest increase in the number within a particular stratum, such as men's outer wear manufacturers.

A random sample may itself be treated as a stratified sample, since it will tend to include a number of firms from each stratum in proportion to the population of that stratum, i.e. the sampling fraction for each stratum will be approximately the same. This fact, of course, is the basis from which Table 4.4 was constructed. However, positive stratification is more desirable, even for calculating the overall mean, since errors are likely to be less than when random selection is relied upon to provide stratification.

As can be seen from the calculations, both the random and systematic samples indicate the mean size of Manchester clothing firms within the acceptable range of ± 15 employees at the 95 per cent level, and the estimates of this particular population parameter provided by the two samples are in close agreement. The original sample of 50 firms (Table 4.1) indicated a mean size which apparently is rather high, although the value lies within the range suggested by the samples of 107 and 118 firms. We may conclude that the mean number of employees per firm in the Manchester clothing industry is of the order of 45–50 but, as shown by the values of the standard deviation, the spread of firm sizes is rather wide. Although mean values for individual strata may not be very significant, Table 4.5 does indicate some difference in size between the various branches of the clothing industry. It is wise, therefore, to investigate this in relation to the proportion of firms within each branch of the industry.

Each subset of firms may be regarded as part of a binomial distribution, in which the proportion of all clothing firms within the subset is p and the proportion of all firms not within the subset is $q = 1 - p$. Alternatively, p and q may be regarded as percentages: then $q = 100 - p$. In order to estimate from a sample the proportion of the total population which occurs within a particular subset, it is necessary to calculate $p \pm z_c \, \sigma_p$, where σ_p, the standard error of the estimate of the proportion is $\sqrt{\dfrac{p \cdot q}{n}}$ when sampling is with

replacement and $\sqrt{\dfrac{p \cdot q}{n}}\sqrt{\dfrac{N-n}{N-1}}$ when samples, once drawn, are not replaced. Of the sample of 107 firms represented in Table 4.2, 36 (33·64 per cent) are in the rainwear category. At the 95 per cent level, therefore, we may estimate that the true proportion of Manchester clothing firms manufacturing rainwear is

$$p \pm z_c \sqrt{\frac{p \cdot q}{n}}\sqrt{\frac{N-n}{N-1}} = 33 \cdot 64 \pm 1 \cdot 96 \sqrt{\frac{33 \cdot 64 \times 66 \cdot 36}{107}}\sqrt{\frac{719}{825}}$$

per cent or $33 \cdot 64 \pm 8 \cdot 36$ per cent. This indicates that between $25 \cdot 28$ per cent and $42 \cdot 00$ per cent of the Manchester firms make rainwear. In order to reduce the standard error whilst retaining the same confidence level, it would be necessary once more to increase the size of the sample. In order to find the proportion to within 5 per cent, we would have to operate with a standard error of $\dfrac{5}{1 \cdot 96}$ per cent.

Thus $\sigma_p = 2 \cdot 55 = \sqrt{\dfrac{33 \cdot 64 \times 66 \cdot 36}{n}}\sqrt{\dfrac{826-n}{825}}$, and $n = 242 \cdot 72$. A

large sample of 243 firms therefore would have to be taken if we were to operate within the 5 per cent limits for the proportion. Clearly, it would not be practicable to carry out such a wide ranging survey, for the time and effort involved would be considerable. The random

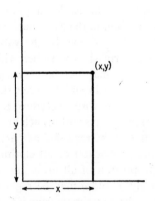

Figure 4.1 Co-ordinates

sample of 107 firms (Table 4.2) included 10 men's outer wear firms, or 9·3 per cent of the total, and the systematic sample of 118 firms (Table 4.3) included 7 men's wear firms, 5·9 per cent of the total. There is little point in estimating these proportions to within ± 5 per cent, but to ensure an accuracy of ±1 per cent would necessitate increasing the samples to 659 and 597 firms so that accurate estimates of the proportion of Manchester clothing firms actually en-

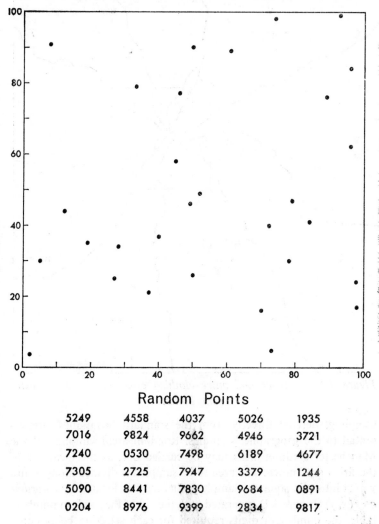

Random Points

5249	4558	4037	5026	1935
7016	9824	9662	4946	3721
7240	0530	7498	6189	4677
7305	2725	7947	3379	1244
5090	8441	7830	9684	0891
0204	8976	9399	2834	9817

Figure 4.2 A random point sample

gaged in the manufacture of men's wear cannot be achieved in practice.

Since geographical data have a spatial element, methods of spatial

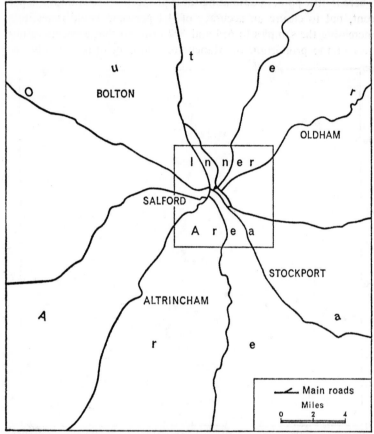

Figure 4.3 The inner and outer clothing areas of the Manchester conurbation

sampling, derived directly from the standard techniques, are essential to the geographer's studies. Random areal sampling of the Manchester clothing firms may be carried out, as the locations of the firms are known. The area to be sampled is divided along x and y axes into grid squares. Sample points may be located by *co-ordinates* (x, y) (Figure 4.1) generated with the aid of the random numbers table, the number of digits required for each selection being equal to the number of figures in the grid reference. Figure 4.2 shows a random point sample using a four-figure reference. Few points are likely to fall exactly on firms as these occupy only a minute proportion of the area of Manchester. To overcome this difficulty, the nearest firm to each point might be included in the sample, although this probably would yield more than 107 firms, as two or more may be equidistant from a point. In fact, different firms may be at the

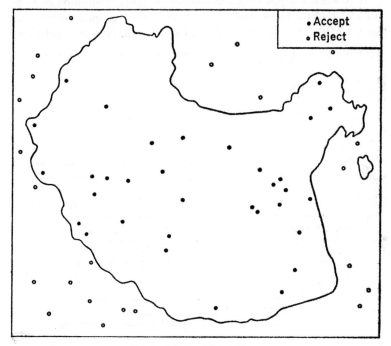

Figure 4.4 A random point sample for an irregular area

same site, as multi-occupation of premises is a common feature in the clothing industry. However, the fact that the sample may include more than 107 firms is of no importance—107 is the minimum size desired and a larger sample merely yields a smaller standard error.

The clothing firms are not evenly distributed throughout Manchester, a large population being concentrated in a 'clothing area' near the city centre, (Figure 4.3). As each point on the map has an equal chance of selection, more points will be selected in the larger outer conurbation than in the smaller 'clothing area'. The proportion of points falling in each area will reflect their relative size. Thus, because they have a greater chance of being included, the firms in the outer area will be over-represented in the sample, which will be biased against firms in the 'clothing area'. As the firms in the outer conurbation may differ in size and in their type of manufacture from those in the centre, a random point sample is unsuitable in this case, an element other than chance (i.e. the spatial distribution of firms) having entered into the selection.

It is when the phenomenon studied is spread continuously over an area that the areal random point method of sampling is particularly useful. Thus, sample values of soil depth, type of vegetation,

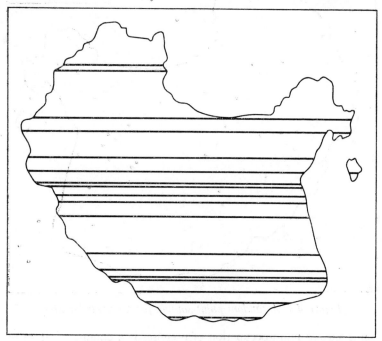

Figure 4.5 A random line sample

etc. may be obtained to provide estimates of mean soil depth, the proportion of the area covered by various vegetation types, etc. A grid somewhat larger than the areal unit studied is necessary to ensure that the selection of random points 'covers' the whole area (Figure 4.4). All points outside the area are discarded.

Two useful variations of the random point method may be employed to provide greater coverage of an area. Instead of points, the sample may be made up of lines or of small areas. Random lines may be generated from one axis (Figure 4.5) and the proportions which lie in different areas (e.g. vegetation types) may be calculated. The total length of line in each type of area (vegetation) must be expressed as a percentage of the total length of all the lines generated. Alternatively, small, equal-size squares to the north-east (or any other direction) of random points may be treated as sample areas. The proportions of the area of all such squares covered by each vegetation type, etc. may be calculated rapidly.

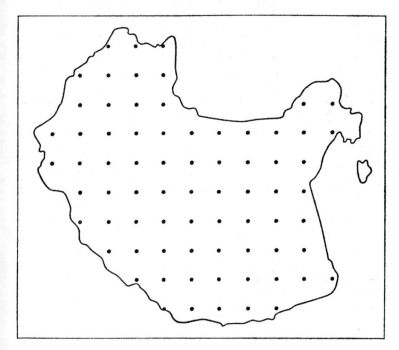

Figure 4.6 A systematic point sample

A systematic areal sample consists of a regularly spaced set of points, lines or squares (Figure 4.6). To overcome some of the disadvantages of systematic sampling, a *systematic random sample* sometimes may be taken (Figure 4.7). In this method one point is taken in each square, but the co-ordinates of the point are generated randomly. Like the random point method, systematic and systematic random point sampling are unsuitable for application to the Manchester clothing firms, but may be used when data are distributed continuously over an area. However, different areas may be regarded as strata, and areal stratified sampling may be used to test for differences between them. The inner and outer Manchester areas thus may be regarded as different strata and a random sample may be taken from each. We know that 208 of the 826 Manchester clothing firms are located within the outer area. Accepting still that 107 firms is the smallest overall sample with which we can work usefully, therefore, an areal stratified sample of 107 firms must consist of 27 firms from the outer area and 80 from the inner area. In practice,

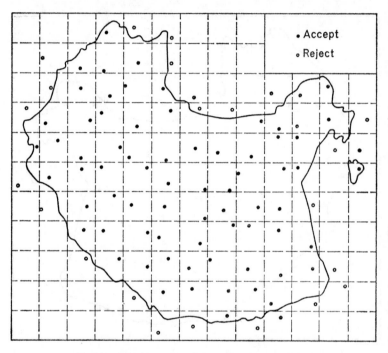

Figure 4.7 A systematic random point sample

areal sampling would not be carried out as we know the location of each of the Manchester clothing firms. By listing separately the firms from the inner and outer areas and drawing a random sample of the required size from each list, we would achieve the same result as if we generated random grid references and took the nearest firm, or firms, to each. It would be simpler to use the lists than to use the map, and the sampling would be completed more rapidly. Having drawn an areally stratified sample, we may treat it in exactly the same manner as the other samples dealt with in this chapter. However, we may like to go further and try to assess whether there is any significant difference between firms from the inner 'clothing area' and those from the rest of the Manchester conurbation. The means by which this can be done are discussed in Chapter 6.

In conclusion, it must be emphasized again that all sampling has to be approached with great caution. The purpose of the exercise must be borne in mind constantly; if it is not, the estimation of parameters may involve much more effort, and much more tedious calculation than is really necessary.

5. Spatial patterns

Many questions can be asked about the spatial distribution of the three sets of data considered in this book. How do the units—people, farms or firms—in the small areas compare with similar units within the larger regions in which they are situated, or with units in the country at large? How are the units arranged spatially? About which point do they cluster? How are the component units scattered about this point? As the data are amenable to measurement, such questions can be answered in precise numerical terms.

Very often a balanced view of a geographical situation may be obtained only if small areas are viewed within the context of their wider environment. Are the farms on Rousay—or Manchester clothing firms or Lancashire boroughs—smaller, larger or of the same size as those of Scotland, Great Britain, etc.? Where complete information is available, simple comparisons of spatially discrete units are easily achieved. Thus, we can state that, with a mean of 4·3 people per acre, Lancashire has a higher population density than has the nation as a whole (0·9 per acre). However, we cannot state whether or not the size of Manchester clothing firms differs from that of all clothing firms, as complete information is not available; the implications of using samples in making spatial comparisons were discussed in Chapter 4.

The importance of the clothing industry in Manchester relative to that of other industries within the conurbation and to the clothing industry elsewhere can be assessed fairly simply. We let the total number of workers in Britain be n_1, the total in North-West England n_2, and the total in the Manchester conurbation n_3; the numbers employed in the clothing industry in each area are c_1, c_2 and c_3 respectively. Clearly, then, of the workers in the Manchester conurbation $\frac{c_3}{n_3} \times 100$ per cent are actually employed in the clothing industry, and Manchester has $\frac{c_3}{c_1} \times 100$ per cent of all the clothing workers in Britain.

A rather more useful method of assessing the importance of the

Manchester clothing industry is to calculate the *location quotient* (L.Q.). This compares Manchester's share of national employment in clothing manufacture with its share of all national employment,

i.e. L.Q. $= \left(\dfrac{c_3}{c_1} \times 100\right) \Big/ \left(\dfrac{n_3}{n_1} \times 100\right) = \dfrac{c_3 n_1}{c_1 n_3}$. Substituting actual

values for c_1, c_3, n_1 and n_3, we obtain a location quotient of

$$\frac{43\ 514 \times 24\ 380\ 000}{365\ 700 \times \ \ 1\ 231\ 250} = 2 \cdot 36.$$

Location quotients can be calculated at other levels; thus, the local location quotient for the Manchester conurbation within the context

of the North-West region is $\dfrac{c_3 n_2}{c_2 n_3} = 1 \cdot 55$. Clearly, if an area has equal

proportions of one particular industry and of all other industries, the value of the location quotient for that industry is unity; larger values indicate the extent to which the industry is localised in the area. Thus, the importance of Manchester as a centre of clothing manufacture can be seen both from the overall location quotient of 2·36 and from its dominance within most branches of the industry (Table 5.1).

Table 5.1

	L.Q. (*in national context*)	Local L.Q. (*in regional context*)
Rainwear	11·5	2·2
Men's outer wear	0·9	1·0
Ladies' outer wear	1·9	1·9
Shirts	2·9	1·3
Dresses	1·8	1·2
All clothing	2·4	1·6

It is clear that men's wear manufacturing within the Manchester conurbation occupies a place similar to that of industry as a whole in both the regional and national contexts, but that other branches of the clothing industry are unusually prominent within the area. The very great dominance of Manchester in Britain's rainwear manufacture is revealed by the L.Q. of 11·5. The North West region is itself a centre for clothing manufacture, but we can see that, even in comparison with this area, Manchester still has a significantly greater share of most branches of the industry.

Naturally, employment in the Manchester area changes with time, and the relative growth or decline of a particular industry, or branch of that industry, may be illustrated by changing values of the location quotient. However, care must be taken not to over-emphasise differences of this kind, since the location quotient is a mean value derived for an area within which considerable differences are likely to occur. Local differences may not be revealed by data available only for large areas. The smaller the area from which data are collected separately, the greater is the likely variation of the location quotient for a particular industry. Changes of the boundaries of areas within which data are collected may invalidate comparisons of location quotients derived from different times. Differences in size of areas of data collection may render meaningless any attempt to compare location quotients for a given industry in different cities, regions, etc.

The location quotient is based on the *Lorenz curve*, which is used to compare an uneven distribution with an even one. (It was used originally in studies of the distribution of incomes throughout the population). The data of Appendix C may be treated with the aid of the Lorenz curve in two ways. The population density of Lancashire can be compared with that of Britain, i.e. a location quotient can be calculated to see whether population is localised in Lancashire. The population location quotient thus is:

$$\frac{\text{Lancashire's percentage of Britain's population}}{\text{Lancashire's percentage of Britain's area}} = 4 \cdot 73$$

i.e. population is much more concentrated in Lancashire than in the country as a whole. In the second place, a full Lorenz curve can be drawn for all the administrative areas of the county in order to assess the extent to which population is unevenly distributed.

A Lorenz curve is drawn on a square graph, the x and y axes having comparable scales. Thus, the x axis may be percentage of population and the y axis percentage of area. If the density of population is perfectly even, the curve will be a straight line sloping at 45° to the horizontal (Figure 5.1). The more uneven the distribution, the more concave will be the curve. A Lorenz curve for the population of Lancashire may be calculated by ranking the administrative areas by density of population and then summing both the area and population with increasing rank order. The least crowded administrative areas of Lancashire are Lunesdale Rural District, with 8224 people in 76 267 acres, and North Lonsdale Rural District, with 16 598 people in 127 448 acres; both have a population density of 0·1 per acre. At the other end of the scale are Bootle C.B., with

82 773 people in 3057 acres (27·1/acre) and Salford C.B. with 155 090 people in 5203 acres (29·8/acre). A Lorenz curve for the 126 divisions of Lancashire (Figure 5.2) reveals that only about 5 per cent of the population live in the least crowded half of the county, whereas 50 per cent occupy the most densely populated 10 per cent.

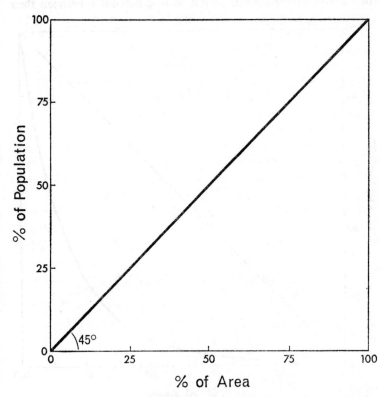

Figure 5.1 The form of a Lorenz curve. The diagonal straight line represents a perfectly even distribution.

More than one Lancashire person in every three lives within the four most densely populated districts—the County Boroughs of Manchester, Liverpool, Bootle and Salford—which together make up slightly more than one twentieth of the county.

The unevenness of a distribution represented by a Lorenz curve can be indicated by expressing the area under the curve as a percentage of that under the perfectly even theoretical distribution. The Lancashire index of 29·9 clearly represents a considerable degree of unevenness. It must be borne in mind, however, that the Lorenz curve is *not* an exact device but merely an approximate visual method of representing a distribution. The total curve is influenced by the size of the individual units considered. For instance, the Rural Districts of Lancashire are generally large and the data published in the Census returns do not permit us to differentiate between their

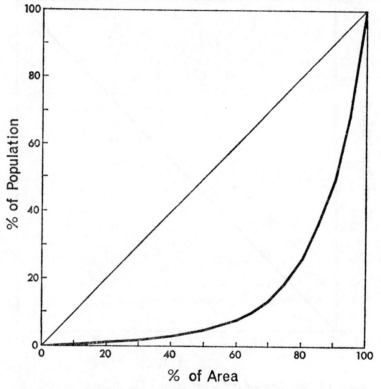

Figure 5.2 Lorenz curve for the distribution of the population of Lancashire

more- and less-densely settled parts. Similarly, the largest urban areas contain zones of very different local population densities which are masked when each County Borough is treated as a single unit.

An alternative method of representing areal distributions is the *topological map*, in which the distance scale of the 'normal' map is replaced by some other one. For example, the map in Figure 5.3(a) shows the administrative divisions of part of east and south Lancashire, and

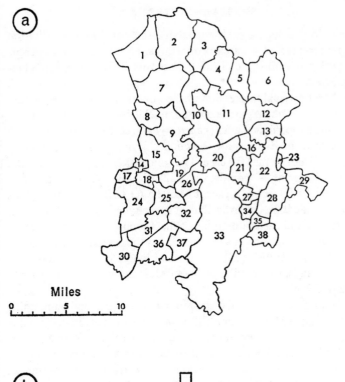

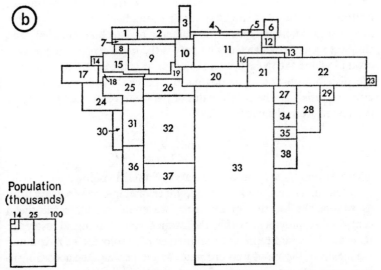

Figure 5.3 (a) *The administrative areas of S.E. Lancashire*
(b) *A topological population map of S.E. Lancashire*

Figure 5.3b is a topological map of the same region, each division being represented by an area proportional to the number of people living within it at the time of the 1961 census. This form of presentation inevitably entails distortion of the real shape of the districts covered, but a visual impression of the population differences within different parts of the region is created. Each of the 38 divisions represented in the topological map is correctly placed with regard to those adjacent to it; no districts are shown as sharing a common boundary where this is not in fact the case. The topological map illustrates the fact that, in terms of population, Salford C.B. (32) is larger than every district adjacent to it except Manchester C.B. (33), although its acreage is little different from that of its neighbours. The districts at the northern and eastern margins of the region included in the maps have lower population densities than have those west of Manchester C.B. This is illustrated by the differences in size of the various units in the two maps, the less densely peopled areas being relatively small in Figure 5.3(b).

The description of spatial distribution patterns of small point units, such as Manchester clothing firms or Rousay dwellings, presents problems different from those associated with large areal units, such as administrative districts. How best can the concentration of Manchester clothing firms be described? Clearly, both the whole set of firms and also each subset should be 'summarised' so that groups involved in different types of manufacturing can be compared. The simplest measure which we can apply is the *mean centre*, the position around which the firms cluster; in effect, it is the centre of gravity of the distribution and is analogous to the arithmetic mean of descriptive statistics. The co-ordinates of the mean centre $(\bar{x}_c, \bar{y}_c)$ may be calculated from the expressions:

$$\bar{x}_c = \frac{\Sigma(x_i P_i)}{\Sigma P_i} \quad \text{and} \quad \bar{y}_c = \frac{\Sigma(y_i P_i)}{\Sigma P_i}$$

where x_i and y_i are the co-ordinates of the ith point and P_i is its population. Thus, if we knew the number of employees in every firm as well as the location of all firms, we could calculate the mean centre of employment within the Manchester clothing industry; in this case P_i would represent the number of employees of a firm and the means of the x and y co-ordinates would be weighted according to firm size. However, as we do not have information about each firm, we can calculate only the mean centre for establishments, i.e. $P_i = 1$; in this case, the means of the co-ordinates are not weighted.

Figure 5.4 shows the mean centres of clothing firms of different types within the Manchester conurbation. All lie within the central area, in which are located three of every four clothing firms of the conurbation.

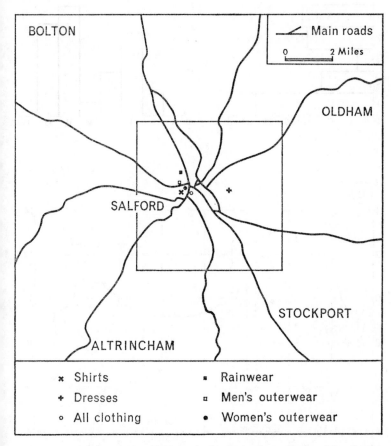

Figure 5.4 The mean centres of the branches of the Manchester clothing industry (by establishment)

Like the arithmetic mean, the mean centre also can be calculated from grouped data. The groups may be grid squares within each of which the total number of units is known; in this case, the centre of each square is assumed to be the mean centre of all the points in that square, just as the mid-interval value is used as the assumed mean for a class in the calculation of the arithmetic mean. In such circumstances, the factor P_i in the expression for the mean centre may

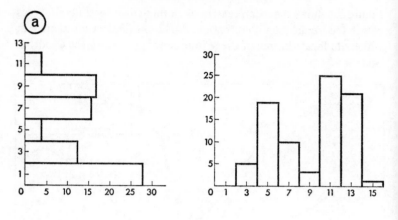

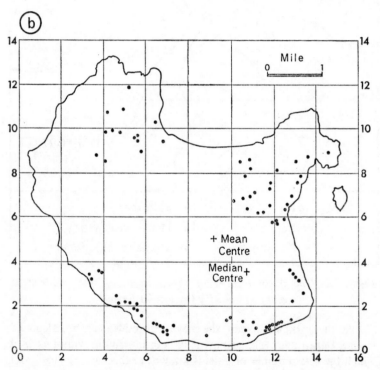

Figure 5.5 (a) *Histograms of the spatial distribution of dwellings on Rousay*
(b) *The spatial distribution of dwellings on Rousay*

represent the number of farms, firms, inhabitants or employees within the ith area, and x_i and y_i are the co-ordinates of the centre of the ith area. The mean centre may be found by regarding the columns based on the x axis, and the rows based on the y axis, as histograms (Figure 5.5(a)). The number of units in each column (f_i) may be multiplied by the mid-interval value of the base of that column (x_i). The value of $\bar{x}_c$ is then found by summing the products $x_i f_i$ and

dividing by the total number of units, n, i.e. $\bar{x}_c = \dfrac{\Sigma x_i f_i}{n}$. Similarly,

$\bar{y}_c = \dfrac{\Sigma y_i f_i}{n}$. In August 1966, there were 84 inhabited dwellings—

caravans and houses—on Rousay. Their mean centre based on a grid with an arbitrary origin lying to the south-west of the island is shown in Figure 5.5(b). The calculation of $\bar{x}_c$ and $\bar{y}_c$ is shown in Table 5.2. If the number of inhabitants of each dwelling was known, the mean centre of population of the island could be calculated in a similar manner, each dwelling being given a weighting factor proportional to the number of inhabitants.

Table 5.2

x_i	f_i	$x_i f_i$	y_i	f_i	$y_i f_i$
1	0	0	1	28	28
3	5	15	3	15	45
5	19	95	5	4	20
7	10	70	7	16	112
9	3	27	9	17	153
11	25	275	11	4	44
13	21	273	13	0	0
15	1	15			

$$\Sigma x_i f_i = 770 \qquad\qquad \Sigma y_i f_i = 402$$

$$\bar{x}_c = \frac{\Sigma x_i f_i}{n} = \frac{770}{84} = 9\cdot17 \qquad \bar{y}_c = \frac{\Sigma y_i f_i}{n} = \frac{402}{84} = 4\cdot79$$

The bimodal frequency distributions in both the x_i columns and the y_i rows reflect the absence of houses in the centre of the island.

As the Rousay dwellings form several separate groups, the mean centre calculated for all of them has little significance, just as an arithmetic mean value may indicate very little about a set of statistics. In a case such as this, therefore, it may be more worth while to calculate separate mean centres for the north-eastern group of dwellings, for the north-western group, etc.

As we can 'average out' a spatial distribution in a similar way to the calculation of 'average values' of arithmetic data, we can determine a spatial equivalent of the median value. The *median centre* $(\bar{x}_m, \bar{y}_m)$ of Rousay dwellings is found simply by drawing the x_i and y_i grid lines which divide the 84 houses, etc., into groups of equal size. There are 42 dwellings west of grid line $\bar{x}_m$ and 42 east of it; there are 42 dwellings to the north and to the south of grid line $\bar{y}_m$ (Figure 5.5(b)). It must be noted that, as the median centre is calculated from an arbitrary origin, its location will change if either the origin or the orientation of the grid lines is changed. Clearly, it is possible to pick out a modal grid square in which there are more dwellings than in any other square. The centre of this square in Figure 5.5(b) has the co-ordinates $x_i = 11$, $y_i = 7$.

Having established a mean centre, we require some measure of the scatter of points about that centre, i.e. a spatial equivalent of the standard deviation. The measure is known as the *standard distance* and, like the mean centre, it can be calculated, in the case of the Manchester clothing industry, either for establishments or for employment and for either grouped or single firms. The expression used is:

Standard Distance (S.D.)

$$= \sqrt{\frac{\Sigma[P_i(x_i - \bar{x}_c)^2]}{\Sigma P_i} + \frac{\Sigma[P_i(y_i - \bar{y}_c)^2]}{\Sigma P_i}}$$

Unfortunately, there is no satisfactory method of representing the standard distance cartographically. The standard distances for the Manchester clothing firms, by establishment, are set out in Table 5.3. We can see that shirt and dress manufacturing are much more dispersed throughout the conurbation than are the other activities. Ladies' outer wear manufacturing is the branch of the industry most concentrated about its mean centre.

Table 5.3

	Standard Distance (km)
Rainwear	4·5
Men's outer wear	4·8
Ladies' outer wear	3·7
Shirts	7·9
Dresses	7·1
All types	6·1

The spatial distribution of establishments such as clothing factories farms or houses is unlikely to be purely random. Almost always

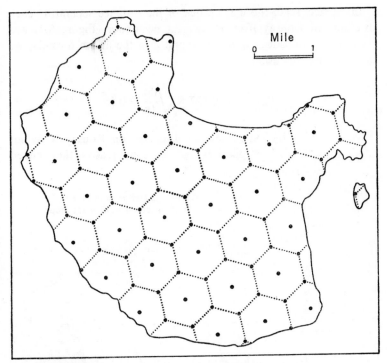

Figure 5.6 A regular hexagonal distribution pattern

factors favouring location within particular zones or areas will have operated. Thus, it may be useful to compare the distribution pattern of point units with that which would exist if the points were distributed in a random manner. We may do this by measuring the distance which separates each point from its nearest neighbour, i.e. that which is closest to it. We then calculate the value R_n, where

$$R_n = 2\bar{d}\sqrt{\frac{n}{A}}$$

$\bar{d}$ is the mean distance between points and their nearest neighbours, A is the area concerned and n is the number of points. This technique is known as *nearest-neighbour analysis*; values of R_n range from zero, when all points are clustered at the same location, to 2·15 when the points have their maximum spacing, in which case they are distributed regularly in a hexagonal pattern (Figure 5.6). If the points are distributed purely at random, the value of R_n is 1.

In August 1966, the mean value of the distance separating each dwelling on Rousay from its nearest neighbour (Figure 5.7) was 0·127 miles. Substituting for n, A and $\bar{d}$ in the above formula, we arrive at the equation

$$R_n = 2 \times 0\cdot127 \times \sqrt{\frac{84}{18}} = 0\cdot55$$

This low value indicates that the dwellings on Rousay were not randomly spaced. It emphasises that, for some reason, they tended

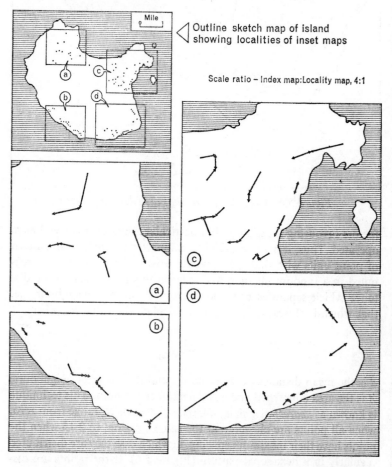

Figure 5.7 The dwellings on Rousay and their nearest neighbours

to be clustered. In fact, the population of Rousay lives almost exclusively around the margins of the island, below the 250 ft contour and within a mile of the sea. Clearly, a number of factors may be responsible for this distribution pattern, some of them physical, some historical, some of them 'accidental', others the result of definite policies or individual decisions. The considerable decrease in the number of inhabited houses on Rousay during the last century has been accompanied by major changes in their pattern of distribution, which has steadily become less dispersed.

6. Contrasts and similarities

The similarities and contrasts between areas have traditionally been the concern of geographers, the description and explanation of areal differentiation being a recurring theme. Before spatial differences in phenomena can be described and explained, however, they must be established, a procedure which often has not received sufficient attention, and in which statistical methods are extremely useful.

It is possible to state categorically and without qualification that in Lancashire the mean size of County Boroughs is greater than that of Municipal Boroughs and Urban Districts. All the relevant facts are available, so a definite statement can be made; no question of probability or statistical significance arises, as it is a certainty that the mean of one population (C.B.s) is greater than that of the other (M.B.s and U.D.s).

If we knew the size of all the Manchester clothing firms, we could also state without reservation whether the mean size of those in the inner 'clothing area' is larger than, equal to, or smaller than that of the firms outside the central area. As we have seen, however, in practice we can only study these firms from samples. Statistical methods enable us to determine whether any difference between sample means is significant, i.e. whether it represents a real difference between the populations or whether the samples are drawn from the same population, the difference in means having occurred by chance.

If we were told the mean number of employees in a sample of firms, we could assign arbitrary numbers to some firms, but the remaining values would be fixed by the mean value provided. For instance, told that the 50 Manchester clothing firms making up Sample 1 of Table 6.1 employed a mean number of 54·56 employees per firm, we could assign arbitrary values to 49 of the firms, but the number of employees to be assigned to the final firm then would be fixed by the mean value of 54·56. The number of items in a set to which arbitrary values can be assigned is known as the number of *degrees of freedom* (d.f.). In the above example, therefore, there are 49 degrees of freedom.

The importance of this concept is perhaps most easily explained in the first instance by taking as an example the Rousay data, where

Table 6.1

Sample 1: Firms from inner area

Class	No. of employees	Class	No. of employees
Rainwear	4	Ladies' outer wear	3
	12		9
	13		14
	15		22
	21		23
	22		25
	24		55
	25		70
	32		200
	43	Shirts	2
	60		3
	66		3
	88		10
	90		16
	100		30
	102		100
	120		150
	150	Dresses	3
	200		5
	510		6
Men's outer wear	6		8
			15
			24
			28
			30
			32
			35
			40
			64

$$\bar{x} = 54\cdot56 \qquad \hat{\sigma}^2 = 6778\cdot46$$
$$\hat{\sigma} = 82\cdot33$$
$$\Sigma x = 2728$$

all the information is available. The rather more difficult situation of the Manchester clothing firms can then be tackled. In a set of N values with a known mean value of $\bar{X}$, there are $(N - 1)$ d.f. In 1966, cattle were kept on 42 Rousay farms. Given that the mean number of cattle per farm was 39·0, we could assign arbitrary numbers of cattle to 41 farms; there are 41 degrees of freedom. If the farms were listed in six groups, each containing seven farms, we could assign arbitrary totals to five groups, but the mean value of 39·0 cattle per farm then would determine the number of cattle to be assigned to the sixth group of farms. For M samples of equal size, there are $(M - 1)$ d.f. *between the samples.* If we were told the mean number

Table 6.1 (continued)

Sample 2: Firms from outer area

Class	No. of employees	Class	No. of employee
Rainwear	6	Dresses	1
	20		1
	50		3
	58		3
	80		6
	130		7
	160		12
	250		14
Men's outer wear	5		16
	18		16
Ladies' outer wear	45		20
Shirts	1		21
	1		25
	3		25
	7		25
	10		30
	30		30
	31		30
	40		55
	52		60
	60		80
	65		165
	90		190
	150		200
	220		
	787		

$$\bar{x} = 68 \cdot 08 \qquad \hat{\sigma}^2 = 14\ 766 \cdot 77$$
$$\Sigma x = 3404 \qquad \hat{\sigma} = 121 \cdot 52$$

of cattle per farm within each of the six groups of farms, it would be possible to assign arbitrary numbers to six of the seven farms in each group. As there are $\dfrac{N}{M}$ variables (farms) in each group, there are $\left(\dfrac{N}{M} - 1\right)$ d.f. *within each sample*. The total number of farms to which arbitrary numbers of cattle may be assigned in this case is $6 \times \left(\dfrac{42}{6} - 1\right) = 36$; there are $M\left(\dfrac{N}{M} - 1\right)$ d.f. *within all samples*.

In a similar manner, it is possible to obtain values for the 'between group', 'within-group' and total ('within-all-groups') sums of squares. Table 6.1 shows the numbers of employees in two samples of Manchester clothing firms, one from the central area, the other from the outer zone. Since in this case $N = 100$ and $M = 2$, there are 99 d.f. in all, with one between the samples, 49 within each sample and 98 within all samples. The mean number of employees per firm in Sample 1 is 54·56 and the sum of the squares of deviations of individual firm sizes from this value is 335 121·11. For Sample 2, the corresponding values are 68·08 and 723 571·68. To obtain the 'within-sample' sum of squares we add together the sums of squares calculated for the individual samples, i.e. 335 121·11 + 723 571·68 = 1 058 692·79. In order to determine the 'between-sample' sum of squares, we assume each firm to be equal in size to the mean value of the sample to which it belongs. Thus, we have 50 firms of size 54·56 and 50 of size 68·08. The squares of deviations from the overall mean size of 61·32 now may be recalculated; this gives us the 'between-sample' sum of squares. Since each firm differs by 6·76 from the mean value, the answer is $100(6·76)^2$, or 4569·76. The relevant degrees of freedom total one less than the number of sample averages, i.e. $(2 - 1) = 1$ d.f. As this is so much less than the 'within-sample' sum of squares, we can see that the individual firms within the samples vary much more than do the means of the separate samples. Comparison of the sums of squares thus does not suggest that there is a significant difference between the two samples of Table 6.1.

The geographer often faces problems of the type considered above. Were two samples of data drawn from the same population, or is there a significant difference between them and therefore between the populations from which they were drawn? Simply comparing 'within-group' and 'between-group' mean square values may not be an adequate way of deciding the matter. In fact, tests on samples cannot prove that they have the same parent group, since samples necessarily are incomplete data; it is always possible that the 'missing' data might be such as to indicate differences in the parent populations of the samples. Analysis of samples can indicate only the probability that differences exist between the populations from which they were drawn.

In order to assess the probability that samples were drawn from the same population, we assume that they were and the assumption is tested. In other words, we assume that there is no difference, the samples being drawn from the same parent group and their different characteristics being no more than chance variations. Such a negative assumption about samples is known as a *null hypothesis*; should it be shown to be wrong, it must be rejected, i.e. we can no longer assume that there is no significant difference between the samples.

If the assumption is not shown to be false, the case is 'not proven'; a null hypothesis cannot be proved correct.

If the difference between samples is one likely to occur by chance once in twenty times, it should not be accepted as significant. When the probability is greater than 5 per cent, a significant difference is not proven. (We cannot say that the difference is not significant, since larger samples might have brought to light differences greater than those displayed by the samples tested.) A difference likely to occur by chance only once in twenty times (5 per cent probability) is probably significant; there is a $19 - 1$ possibility that the samples came from different parent groups. Where the difference is to be expected as a result of chance only once in 100 times (1 per cent probability), it is very unlikely that the assumption is correct; the difference is significant. Thus, two or more sets of data may be compared by determining the probability that the observed differences are only chance variations, arising naturally in the course of taking different random samples from the same parent population.

The difference between samples may be indicated by the standard deviation of the distribution of the differences between their means, known as the *standard error of the difference* (in the same way as the standard deviation of the distribution of the sample means is termed the standard error of the mean). The standard error of the difference is thus equal to the square root of the variance of the distribution of the differences between sample means. Now the variance of the sum or difference of two or more independent random variables is equal to the sum of the variances of the individual random variables:

$$\text{Var}_{(x_1 - x_2)} = \sigma_1{}^2 + \sigma_2{}^2$$

and the variance of the sum or difference of independent random sample means is

$$\text{Var}_{(x_1 \pm x_2)} = \sigma_{\bar{x}_1}^2 + \sigma_{\bar{x}_2}^2 = \frac{s_1{}^2}{n_1} + \frac{s_2{}^2}{n_2}$$

and the standard error of the difference,

$$\text{S.E.}_{(\bar{x}_1 - \bar{x}_2)} = \sqrt{\frac{s_1{}^2}{n_1} + \frac{s_2{}^2}{n_2}}$$

However, since we must have an unbiased estimate of the population variance when comparing samples, it is necessary to apply

Bessel's correction (p. 36) to the sample variance. Thus, it follows that, more correctly,

$$\text{S.E.}_{(\bar{x}_1 - \bar{x}_2)} = \sqrt{\frac{s_1^2 n_1}{n_1(n_1 - 1)} + \frac{s_2^2 n_2}{n_2(n_2 - 1)}}$$

and, for the samples of Table 6.1.

$$\text{S.E.}_{(\bar{x}_1 - \bar{x}_2)} = \sqrt{\frac{335\ 121 \cdot 11}{49} + \frac{723571 \cdot 68}{49}}$$

$$= \frac{1}{7}\sqrt{1\ 058\ 692 \cdot 79}$$

$$= 146 \cdot 99$$

Clearly, as each sample was drawn from a separate area, some variation must arise from the random selection of individual firms within a particular population, but other differences may represent real differences between the populations of firms in the two areas, rather than the operation of chance. The significance of the difference between samples is assessed by comparing the difference between their means with the standard error of the difference, a test known as *Student's test*. The difference between the sample means of Table 6.1 is 13.52 and the ratio

$$t = \frac{\text{Difference between sample means}}{\text{Standard error of the difference}} = \frac{(\bar{x}_1 - \bar{x}_2)}{\text{S.E.}_{(\bar{x}_1 - \bar{x}_2)}}$$

$$= \frac{13 \cdot 52}{146 \cdot 99} = 0 \cdot 09.$$

The larger are values of t, the lower is the probability that an assumption of no significant difference is correct. With 98 degrees of freedom, chance variations would cause t to exceed $2 \cdot 62$ only once in about 100 times (Appendix F), so $t = 2 \cdot 62$ would indicate the difference between the means of two randomly selected samples to be statistically significant; a real difference between the populations from which the samples were drawn would be indicated. However, $t = 0 \cdot 09$ is a result which could arise by chance very often when there is one degree of freedom. Since this probability exceeds 5 per cent, a significant difference between the sample means of Table 6.1 is not proven, and we cannot reject the hypothesis of no significant difference between the populations of firms in the two areas sampled.

An alternative method of dealing with the two samples of Manchester clothing firms is the use of the *analysis of variance*, which assesses the contribution made by each separate factor to the total variability of a set of data. We may assume that the two equal-sized samples of firms detailed in Table 6.1 are drawn from a single population, and that, in terms of employment, there is no significant difference between them. This null hypothesis may be tested by comparing the 'within-sample' and 'between-sample' variances since, if the assumption is correct, both these should be estimates of, or approximations to, the variance of the single population from which the different samples were drawn. A sample variance may be calculated by dividing the sum of squares by the number of items within the sample. With small samples, however, a better estimate of the population variance is obtained by dividing the sum of squares by the number of degrees of freedom, $(n - 1)$. We may calculate the 'between-sample' estimate of variance and the 'within-sample' estimate of variance in a similar manner. Thus, for the firms of Table 6.1,

$$\text{'Between-sample' variance} = \frac{4569 \cdot 76}{1} = 4569 \cdot 76$$

$$\text{and 'within-sample' variance} = \frac{1\ 058\ 692 \cdot 79}{98} = 10\ 802 \cdot 99$$

$$\text{The ratio, } F = \frac{\text{'between-sample' variance}}{\text{'within-sample' variance}} = 0 \cdot 42.$$

The significance of the *variance ratio* at different confidence levels (probability levels) may be checked by reference to tabulated values, a test commonly termed *Snedecor's F test* (Appendix H). For 98 d.f. within all samples and 1 d.f. between the samples, $F = 0 \cdot 42$ is not significant at the 5 per cent confidence level. The null hypothesis therefore is not shown to be false, and we must assume that the two samples probably are not drawn from two different populations.

Having applied a variety of tests to them, we have had to conclude that a significant difference in the sizes of clothing firms represented by the two samples of Table 6.1 is not proven. However, there remains the possibility that the firms in the inner and outer areas differ in terms of the proportions engaged in various types of manufacture. Are they different and are the differences significant? The samples show the following proportions:

	Rainwear	Men's outer wear	Ladies' outer wear	Shirts	Dresses
Sample 1	20	1	9	8	12
Sample 2	8	2	1	15	24

The apparent differences can be tested by means of the χ^2 (chi-squared) test. It is assumed that both samples are drawn from a single population which, for each type of manufacturing, has a frequency mid-way between those of the two samples (i.e. the null hypothesis is one of no difference). In Sample 1, 20 of the firms manufacture rainwear; in Sample 2, 8 firms. The expected frequency (E) is $\dfrac{20 + 8}{2} = 14$ firms, whereas the observed frequencies (O) are 20 and 8 firms. For every type of manufacturing in the two samples the observed frequency (O) and expected frequency (E) may be compared. The value $\dfrac{(O - E)^2}{E}$ can be calculated for each type of firm in each sample and the results summed to give a value

$$\chi^2 = \sum \frac{(O - E)^2}{E}$$

The calculation is set out below:

	O		E	$O - E$	$(O - E)^2$	$\dfrac{(O - E)^2}{E}$
	Sample 1	Sample 2	(For both Sample 1 and Sample 2)			
Rainwear	20	8	14	6	36	2·57
Men's outer wear	1	2	1·5	0·5	0·25	0·17
Ladies' outer wear	9	1	5	4	16	3·2
Shirts	8	15	11·5	3·5	12·25	1·07
Dresses	12	24	18	6	36	2

$$\sum \frac{(O - E)^2}{E} = 9$$

Thus for both samples $\chi^2 = 16$.

There are $(n - 1)$ d.f., where n is the number of pairs, i.e. $n = 5$ and there are 4 d.f.

If the expected and actual values all are equal, $\chi^2 = 0$. The value of χ^2 thus indicates the departure of observed values from expected ones. Chance variations in random sampling may be responsible for small departures of χ^2 from a zero value. Larger values, however, indicate a significant difference between the observed and expected values. The significance of the difference may be checked for the appropriate degrees of freedom at different levels of confidence by the use of χ^2 tables (Appendix G). The result for the Manchester firms indicates that, at the 1 per cent level, a significant difference between the two areas is established, the calculated value of χ^2 (16) being greater than the tabulated value (13·277).

As we can see, the χ^2 test enables us to compare frequency distributions rather than mean values. Thus, it could also have been used to test whether the two samples differ significantly in numbers of employees per firm; the samples would have had to be arranged in comparable size classes and the frequency distributions tabulated. The χ^2 test has the advantage over the t test that the whole range of values of the distribution is considered, and not merely the mean and standard deviation.

This approach may be extended to compare the frequency distribution of data in a sample with that which would be expected if the data conformed to some theoretical distribution, such as the normal or the binomial. The departure of the observed frequencies from the theoretical ones gives a measure of the *goodness of fit* with the theoretical pattern. If the observed distribution of data approximates to the assumed pattern, there should be no significant difference between the expected and actual frequencies.

The data of Figure 3.5 may be tested for goodness of fit with the log-normal distribution. For each actual frequency (O) of a given number of employees per firm, there is an expected one (E), derived from the log-normal curve. The goodness of fit is represented by the value of $\dfrac{(O - E)^2}{E}$ and that for the entire distribution by

$$\sum \frac{(O - E)^2}{E} = \chi^2$$

At the 1 per cent level a significant difference between the frequency distribution of the firms and a log-normal distribution is not proven.

Thus, we see that the size distribution of Manchester clothing firms may be log-normal. As far as size is concerned, it seems likely that firms in the inner 'clothing area' do not differ significantly from those outside, but it is likely that the distribution by type differs. It seems that clothing firms nearer the centre of Manchester are more likely to concentrate on the manufacture of rainwear and ladies' wear than are those further out, which may be more concerned with making shirts and dresses.

Differences between samples may be examined in a variety of ways. To assess whether such differences represent real differences of the populations from which the samples were drawn is an important task for the geographer to tackle. The method he will use must depend upon the nature of the data considered, the time available for calculation and the sort of difference which is being considered.

7. Geographic relationships

Central to the study of geography is the evaluation of the relationships which exist between spatially distributed phenomena. Often the geographer has two apparently related sets of numerical information in which items may be paired, one item from each set forming a pair and each of the pairs referring to a particular place or to a particular moment in time. This chapter is concerned with the methods used, and the problems arising, in assessing the presence and significance of apparent relationships between variables.

At first sight, the arable acreage and number of cattle for each holding on Rousay appear to be related. The two values for each holding, which constitute what is known as a *bivariate distribution*, can be plotted graphically on a *scatter diagram* (Figure 7.1), the arable acreage of each unit being shown along the horizontal, or x, axis and the corresponding number of cattle along the vertical, or y, axis. Although there is no *functional relationship* between x and y (i.e. for any one value of x there is not one precise corresponding value of y), it can be seen that the variables increase together in a general way, the points being located in quite a narrow belt on the diagram. The two variables are said to be *directly or positively correlated*, as an increase in the value of one is accompanied by an increase in the value of the other. (If an increase in one were accompanied by a decrease in the other, the variables would be described as *inversely or negatively correlated*; the belt of points would slope in the opposite direction to that shown in Figure 7.1.)

The numerical relationship between the values of x and y does not necessarily mean that the variation of one *causes* the variation of the other. We should not necessarily conclude that the number of cattle possessed by a farmer is fixed by the acreage of his arable land, or that his arable acreage is a response to the number of cattle which need to be fed, although either or both of these conclusions which can be inferred from the data could contain some element of truth. It is possible that both the variables depend on another factor, such as the amount of lowland on each holding; in other words, x and y may be *co-variables*, both of which vary with some third factor. We must be very cautious when inferring relationships; completely

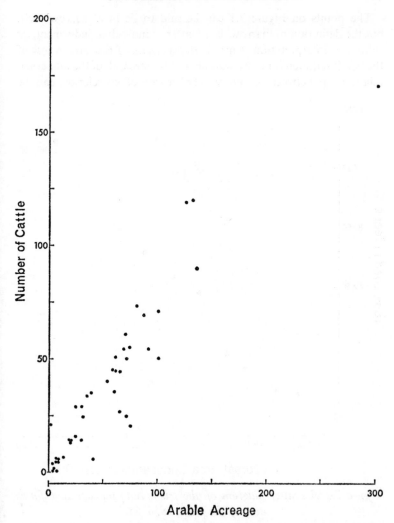

Figure 7.1 A positively correlated bivariate distribution. Arable acreages and numbers of cattle on Rousay farms

spurious correlations may be deduced from a scatter diagram of two totally unrelated variables which increase or decrease over a period of time. The total trunk road mileage in Great Britain in 1956–65, for example, will correlate with the forest area for the same years as both increased progressively during the period (Figure 7.2), but to conclude that one depends upon the other clearly is nonsense.

The points on Figure 7.1 can be said to lie in a 'narrow' belt, but the definition of 'narrow' is a matter of individual judgement, or subjective interpretation. An *objective* measure of the narrowness of the belt is required for any assessment of the strength of the numerical relationship between x and y. The degree of correlation can be

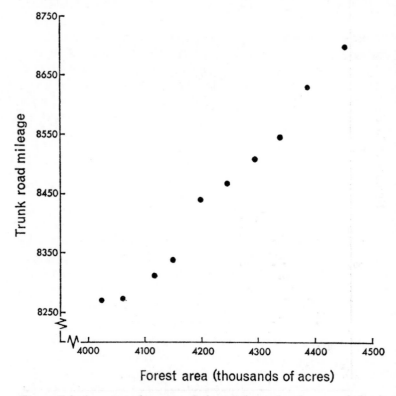

Figure 7.2 A scatter diagram of the trunk road mileage and forest area of the U.K., 1956–65

measured by the use of a numerical *coefficient of correlation*, known as R_{xy}. All correlation coefficients have values between 1 and -1, those near unity indicating a high degree of correlation. However, where R_{xy} has been calculated from a few pairs, there is a high probability that the values of x and y have correlated by chance. Low values of R_{xy} indicate a low degree of correlation and where R_{xy} is equal to zero there is a complete absence of correlation. The coefficient shows how close is the approach to a straight line functional relationship, for which it is equal to 1 or −1, depending on whether the correlation is direct or inverse. However, the coefficient does not provide a quantitative measure of the ratio of one variable to the other.

The significance of a correlation coefficient cannot be assessed until all known factors have been taken into account and the possibility that unknown factors exist has been considered. In most situations with which the geographer is concerned, elements of chance or vagaries of human behaviour operate to disturb any correlation between variables. If the influence of these uncontrolled factors is great, the value of the calculated correlation coefficient will be low and any real relationship between the two variables may be obscured. However, if the coefficient has a high value, we may assume that, despite chance elements, the correlation has real significance. The statistical significance of a correlation can be assessed by using Student's t test, which was mentioned on page 69.

The coefficient of correlation is calculated on the basis of deviations of individual values of x and y from their mean values. By analogy with mechanics, these deviations, d_x and d_y, are known as the first moments about the mean, and their product, $d_x d_y$, therefore is called the *product-moment*. The mean value of $d_x d_y$ is known as the *covariance*, σ_{xy} and the term *product-moment coefficient of correlation* is applied to R_{xy}, since

$$R_{xy} = \frac{\sigma_{xy}}{\sigma_x \sigma_y}$$

The coefficient of correlation between arable acreage and the number of cattle on Rousay may be calculated in the way shown on page 78. Values of x and y are summed and their mean values, $\bar{x}$ and $\bar{y}$, calculated. The deviations of x and y from their respective means are squared to give d_x^2 and d_y^2, from which the standard deviations of x and y can be calculated.

Thus,
$$\sigma_x = \sqrt{\frac{\Sigma d_x^2}{n}} = 52 \cdot 34$$

and
$$\sigma_y = \sqrt{\frac{\Sigma d_y^2}{n}} = 35 \cdot 58$$

The mean value of $d_x d_y$ is
$$\sigma_{xy} = \frac{\Sigma d_x d_y}{n} = 1735 \cdot 41$$

and the coefficient of correlation is

$$R_{xy} = \frac{\sigma_{xy}}{\sigma_x \sigma_y} = 0.93$$

x	y	d_x	d_y	$d_x{}^2$	$d_y{}^2$	$d_x d_y$
4	3	−50·25	−34·25	2525·06	1173·06	1721·06
4	0	−50·25	−37·25	2525·06	1387·56	1871·81
4	3	−50·25	−34·25	2525·06	1173·06	1721·06
5	1	−49·25	−36·25	2425·56	1314·06	1785·31
6	4	−48·25	−33·25	2328·06	1105·56	1604·31
8	4	−46·25	−33·25	2139·06	1105·56	1537·81
8	6	−46·25	−31·25	2139·06	976·56	1445·31
7	6	−47·25	−31·25	2232·56	976·56	1476·56
7	0	−47·25	−37·25	2232·56	1387·56	1760·06
21	13	−33·25	−24·25	1105·56	588·06	806·31
20	12	−34·25	−25·25	1173·06	637·56	864·81
30	14	−24·25	−23·25	588·06	540·56	563·81
25	27	−29·25	−10·25	855·56	105·06	299·81
2	21	−52·25	−16·25	2730·06	264·06	849·06
35	33	−19·25	−4·25	370·56	18·06	81·81
31	24	−23·25	−13·25	540·56	175·56	308·06
19	14	−35·25	−23·25	1242·56	540·56	819·56
54	40	−0·25	2·75	0·06	7·56	−0·68
30	27	−24·25	−10·25	588·06	105·06	248·56
61	51	6·75	13·75	45·56	189·06	92·81
24	16	−30·25	−21·25	915·06	451·56	642·81
65	44	10·75	6·75	115·56	45·56	72·56
65	27	10·75	−10·25	115·56	105·06	−110·18
72	50	17·75	12·75	315·56	162·56	226·31
75	20	20·75	−17·25	430·56	297·56	−357·93
42	6	−12·25	−31·25	150·06	976·56	382·81
67	53	12·75	15·75	162·56	248·06	200·81
14	7	−40·25	−30·25	1620·06	915·06	1217·56
70	61	15·75	23·75	248·06	564·06	374·06
60	36	5·75	−1·25	33·06	1·56	−7·18
71	24	16·75	−13·25	280·56	175·56	−221·93
73	55	18·75	17·75	351·56	315·06	332·81
58	45	3·75	7·75	14·06	60·06	29·06
62	44	7·75	6·75	60·06	45·56	52·31
39	35	−15·25	−2·25	232·56	5·06	34·31
135	90	80·75	52·75	6520·56	2782·56	4259·56
100	71	45·75	33·75	2093·06	1139·06	1544·06
86	68	31·75	30·75	1008·06	945·56	976·31
91	53	36·75	15·75	1350·56	248·06	578·81
80	73	25·75	35·75	663·06	1278·06	920·56
100	50	45·75	12·75	2093·06	162·56	583·31
125	118	70·75	80·75	5005·56	6520·56	5713·06
132	120	77·75	82·75	6045·06	6847·56	6433·81
300	170	245·75	132·75	60 393·06	17 622·56	32 623·31
				$\Sigma d_x{}^2$	$\Sigma d_y{}^2$	$\Sigma d_x d_y$
				= 120 528·3	= 55 686·25	= 76 358·25

This indicates a high degree of correlation between arable acreage and the number of cattle.

The value of Student's t may be calculated from the equation

$$t = \frac{R_{xy}\sqrt{n-2}}{\sqrt{1-R_{xy}^{2}}},$$

where n is the number of pairs and the degrees of freedom are $(n-2)$; R_{xy} is always taken as positive, merely for convenience. For the data under consideration, $t = \dfrac{(0.93)\sqrt{42}}{\sqrt{1-(0.93)^{2}}} = 16.41$. Student's

t test indicates that the correlation between cattle and arable acreage is significant at the 99 per cent level, i.e. the probability that the relationship results from chance is less than 1 in 100.

Since many pairs (44) are involved in this correlation, it is not surprising that a correlation coefficient as high as 0·93 indicates a significant relationship. Indeed, a much lower value of R_{xy} would give a fairly large value of t. It may be of interest, therefore, to determine the minimum value of R_{xy} which the t test indicates to be significant at the 99 per cent level. The t tables show that, with 40 degrees of freedom, the critical value of t at the 99 per cent level is 2·704. Substituting in the expression for t, we have

$$2.704 = \frac{R_{xy}\sqrt{42}}{\sqrt{1-R_{xy}^{2}}}$$

which gives $R_{xy} = 0.39$.

Thus, a correlation coefficient as low as 0·39 would suggest a significant relationship between the arable acreage and number of cattle on a group of 44 farms.

If the correlation coefficient, R_{xy}, is squared and multiplied by 100, it gives a value known as the *coefficient of determination*, which indicates what percentage of the variation in one variable is explained by variation in the other. In the case under discussion, the coefficient of determination is about 88, so that we may conclude that about 88 per cent of the variation of arable acreage from farm to farm in Rousay is 'explained' by the variation of the number of cattle, and vice versa.

Sometimes, the actual numerical values of x and y are not available, but only their relative order of magnitude. In such cases, the *ranks*, the positions of the variables in numerical order, are used to calculate a *coefficient of rank correlation*, ρ. This is also of use where a rapid estimate of the strength of a correlation is required. The coefficient of rank correlation is the product-moment coefficient of correlation calculated from the ranks of x and y instead of from their absolute values. It requires less arithmetic than a full product-moment calculation, and results in an approximation to R_{xy}.

The values of a variable may be ranked from the highest to the lowest, or vice versa. For example:

$$17 \quad 15 \quad 12 \quad 3 \quad 2 \quad 1$$

may be ranked as

1	2	3	4	5	6	$\Sigma R = 21$

or as

6	5	4	3	2	1	$\Sigma R = 21$

Where two or more of the values of a variable are equal (or *tied*) they are given the same rank. For example:

$$17 \quad 15 \quad 15 \quad 3 \quad 2 \quad 1$$

would be ranked as

1	2·5	2·5	4	5	6	$\Sigma R = 21$

or as

6	4·5	4·5	3	2	1	$\Sigma R = 21$

The sum of the ranks (21) therefore is the same as in the first example.

The fact that the sum of n ranks always equals the sum of the first n natural numbers makes the calculation of ρ relatively simple. The coefficient of rank correlation was introduced by Spearman, who showed that

$$\rho = 1 - \frac{6\Sigma\, D^2}{n(n^2 - 1)}$$

where D is the rank difference and n is the number of pairs. It should be noted that, if all units in one set have equal rank, then ρ is always 0·5. The coefficient is therefore unsuitable for cases where more than a few values are tied. Tied values cause D^2 to be too low. This can be overcome by adding a correction factor T, equal to $\sum\left(\dfrac{i^3 - i}{12}\right)t_i$, where i is the number of values in a set of tied values and t_i is the number of ties involving i values. Thus

$$T = \tfrac{1}{2}t_2 + 2t_3 + 5t_4 \ldots \text{etc.}$$

The calculation of the product-moment coefficient of correlation was a lengthy, time-consuming operation. On the other hand ρ can be calculated relatively quickly although it does not use the data to the full. The calculation is set out opposite.

x (Arable acres)	y (Cattle)	R_x	R_y	$D(R_x - R_y)$	D^2
4	0	42	43·5	−1·5	2·25
4	3	42	40·5	−1·5	2·25
4	3	42	40·5	−1·5	2.25
5	1	40	42	−2	4
6	4	39	38·5	0·5	0·25
8	4	35·5	38·5	−3	9
8	6	35·5	36	−0·5	0·25
7	0	37·5	43·5	−6·0	36
7	6	37·5	36	1	1
21	13	31	32	−1	1
20	12	32	33	−1	1
30	14	27·5	30·5	−3	9
25	27	29	23	6	36
2	21	44	27	17	289
35	33	25	21	4	16
31	24	26	25·5	0·5	0·25
19	14	33	30·5	2·5	6·25
54	40	22	18	4	16
30	27	27·5	23	4·5	20·25
61	51	19	12	7	49
24	16	30	29	1	1
65	44	16·5	16·5	0	0
65	27	16·5	23	−6·5	42·25
72	50	12	13·5	−1·5	2·25
75	20	10	28	−18	324
42	6	23	36	−13	169
67	53	15	10·5	4·5	20·25
14	7	34	34	0	0
70	61	14	8	6	36
60	36	20	19	1	1
71	24	13	25·5	−12·5	56·5
73	55	11	9	2	4
58	45	21	15	6	36
62	44	18	16·5	1·5	2·25
39	35	24	20	4	16
135	90	2	4	−2	4
100	71	5·5	6	−0·5	0·25
86	68	8	7	1	1
91	53	7	10·5	−3·5	12·25
80	73	9	5	4	16
100	50	5·5	13·5	−8	64
125	118	4	3	1	1
132	120	3	2	1	1
300	170	1	1	0	0

$$\Sigma D^2 = 1523·75$$

The values of x and y are ranked as R_x and R_y respectively. The differences between R_x and R_y are calculated, squared and summed. Substituting in Spearman's formula,

$$\rho = 1 - \frac{6 \Sigma D^2}{n(n^2 - 1)} = 0·89$$

In this case, the effect of tied values can be neglected. The number of ties involving two values ($i = 2$) is 5 in column R_x and 8 in column R_y; thus $t_2 = 13$. The number of ties involving 3 values ($i = 3$) is one in column R_x and 2 in column R_y; $t_3 = 3$. There are no ties involving

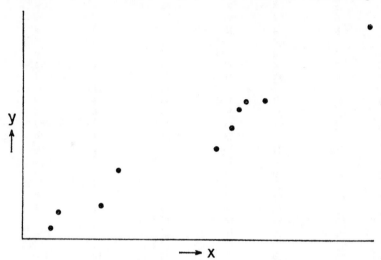

Figure 7.3 A hypothetical bivariate distribution in which there is perfect correlation by ranks but imperfect correlation by absolute values

more than 3 values. Thus $T = \frac{1}{2}.13 + 2.3 = 12\cdot5$. Adding this to $\Sigma\,D^2$ gives $\Sigma\,D^2 + T = 1536\cdot25$. Then,

$$\rho = 1 - \frac{6(\Sigma\,D^2 + T)}{n(n^2 - 1)} = 0\cdot89$$

For the same two variables, the value of R_{xy} is $0\cdot93$. Although values of R_{xy} and ρ may differ slightly, ρ like R_{xy}, may be tested for significance with the use of the t distribution, provided there are more than 10 pairs of values. In an extreme case, the rank coefficient may be equal to unity for data which are not perfectly correlated by product-moments. (A hypothetical example is shown in Figure 7.3.) Valuable though it is, therefore, the use of ranks rather than actual values must be treated with caution.

The high degree of rank correlation for arable acreage and cattle numbers on Rousay holdings indicates that the two variables are related, but ρ is not a measure of the real numerical relationship. Figure 7.1, the scatter diagram of the values of x and y, demonstrates the real numerical relationship. As a few of the values in this bivariate distribution are much larger than the others, many points have to be crowded together in order to fit all into a diagram of

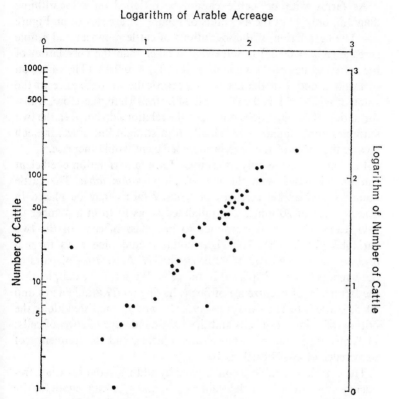

Figure 7.4 The logarithms of the arable acreages and numbers of cattle on Rousay farms plotted as a bivariate distribution

reasonable size. In Chapter 3 it was shown that the use of a logarithmic scale overcomes the problem of skewness in a simple distribution. Logarithmic scales also are of use with bivariate distributions, since they 'spread out' the points. Figure 7.4 is a scatter diagram of the logarithms of the values of x and y shown in Figure 7.1; logarithmic scales have been used for both sets of data as each set has a few high values and many low ones. In other cases, logarithmic transformation of either or both variables in a bivariate distribution may be desirable in order to convert a curvilinear relationship into a more manageable linear one.

As farms with no cattle cannot be included on a logarithmic diagram, only 42 of the 44 points of Figure 7.1 are shown on Figure 7.4. The correlation of the logarithms of cattle numbers and arable acreage ($R_{xy} = 0\cdot865$) is somewhat weaker than the correlation of the actual values of the same 42 pairs ($R_{xy} = 0\cdot929$). (The omission of farms 2 and 9 in the list makes practically no difference to the value of R_{xy}, which is $0\cdot932$ for all 44 farms.) Here, therefore, transformation of the data does not make the relationship between the two variables approximate more closely to a straight line since, in such a case, the value of the correlation coefficient would increase.

For data available only in grouped form, a correlation coefficient can be calculated with the aid of a *correlation table*. The cattle numbers and arable acreages of Rousay farms may be placed in classes, each of 20 units, and tabulated so as to form a number of *cells*, each of which corresponds to one class of each of the two variables (Table 7.1). The eight columns and nine rows form a *bivariate frequency table* of 72 cells; the *cell frequencies* of pairs of occurrences (f_c) are displayed in the cells. We see, for example, that the frequency of occurrence of farms having 60–79 arable acres and 40–59 cattle is 6. The *class frequencies* (f_x and f_y) are listed along the bottom of columns of cells and the right-hand edge of rows of cells; 11 farms are in the class 60–79 arable acres, and the frequency of occurrence of 40–59 cattle is 10.

The correlation table is constructed by adding to the bivariate frequency table. In order to determine σ_x, σ_y and σ_{xy} mean values for the grouped data are assumed and d_x and d_y are calculated for each row and column of the bivariate table. The results are placed above the appropriate column and to the left of the appropriate row, and values of f_x, $f_x d_x$, $f_x d_x^2$, f_y, $f_y d_y$ and $f_y d_y^2$ are then calculated and inserted around the table. For each cell, the product $f_c d_x d_y$ is calculated and placed in the lower right corner. Products are summed by column and by row to give the grand total, $\Sigma f_c d_x d_y$; a check of the results is given automatically by summing in both ways.

Table 7.1 is based on assumed mean values of $\bar{x}_0 = 49\cdot5$ acres and $\bar{y}_0 = 29\cdot5$ cattle. For each column, d_x is found from the expression

$$d_x = \frac{x_0 - \bar{x}_0}{c_x}$$

Table 7.1 Correlation Table

Arable Acreage

Cattle

y_0	d_x / d_y	x_0 9·5	29·5	49·5	69·5	89·5	109·5	129·5	309·5	f_y	$f_y d_y$	$f_y d_y^2$	
		−2	−1	0	1	2	3	4	13				
169·5	7								1 [91]	1	7	49	91
149·5	6									0	0	0	0
129·5	5							1 [20]		1	5	25	20
109·5	4							1 [16]		1	4	16	16
89·5	3							1 [12]		1	3	9	12
69·5	2				1 [2]	2 [8]	1 [6]			4	8	16	16
49·5	1			2 [0]	6 [6]	1 [2]	1 [3]			10	10	10	11
29·5	0	1 [0]	5 [0]		4 [0]					10	0	0	0
9·5	−1	11 [22]	4 [4]	1 [0]						16	−16	16	26
f_x		12	9	3	11	3	2	3	1	44	21	141	192
$f_x d_x$		−24	−9	0	11	6	6	12	13	15			
$f_x d_x^2$		48	9	0	11	12	18	48	169	315			
		22	4	0	8	10	9	48	91	192			

The bivariate frequency table is enclosed by the heavy line. The right-hand column and bottom row show the sum of the numbers in the lower right corners of cells.

(N.B. Totally blank columns have been excluded from the bivariate frequency table between $x_0 = 129·5$ and $x_0 = 309·5$.)

and for each row

$$d_y = \frac{y_0 - \bar{y}_0}{c_y}$$

where x_0 and y_0 are class mid-interval values and c_x and c_y are the class intervals. The remaining items at the bottom and left hand side

D

of the correlation table then are found by simple multiplication. The correlation coefficient for the grouped data from Rousay is

$$R_{xy} = \frac{\sigma_{xy}}{\sigma_x \sigma_y},$$

where

$$\sigma_{xy} = c_x c_y \left[\frac{\Sigma f_c d_x d_y}{n} - \left(\frac{\Sigma f_x d_x}{n} \right) \left(\frac{\Sigma f_y d_y}{n} \right) \right]$$

$$\sigma_x = c_x \sqrt{ \frac{\Sigma f_x d_x^2}{n} - \left(\frac{\Sigma f_x d_x}{n} \right)^2 }$$

and

$$\sigma_y = c_y \sqrt{ \frac{\Sigma f_y d_y^2}{n} - \left(\frac{\Sigma f_y d_y}{n} \right)^2 }$$

Thus,

$$R_{xy} = \frac{20.20 \left[\frac{192}{44} - \left(\frac{15}{44} \right) \left(\frac{21}{44} \right) \right]}{\left[20 \sqrt{ \frac{315}{44} - \left(\frac{15}{44} \right)^2 } \right] \left[20 \sqrt{ \frac{141}{44} - \left(\frac{21}{44} \right)^2 } \right]}$$

$$= \frac{400(4 \cdot 2009)}{400 \sqrt{(7 \cdot 0428)(2 \cdot 9768)}}$$

$$= 0 \cdot 92$$

Thus, the analysis of the grouped data again shows a strong correlation between arable acreages and the numbers of cattle on Rousay farms. The fundamental problem of what this relationship means, however, is still left open. The geographer has to decide whether the amount of land given over to cultivation is determined by the number of cattle on a holding, or whether the amount of arable land is fixed or limited in some way, the number of cattle which a holding can support being dependent on this. In both cases, it could be argued that the relationship exists because the cattle are fed on fodder produced on the holding on which they are reared, which may be a result of the high cost of transporting animal feed from the mainland. The geographer therefore has to formulate his hypothesis using all the available facts. A statistical relationship is only part of the picture, but as an aid to analysis it is of great value in that it states the problem in precise terms. The statistical relationship is a question rather than an answer, a point which cannot be over-emphasized.

The coefficient of correlation is a very useful measure, but it does not tell us whether a change in x is accompanied by a larger or smaller change in y; it merely indicates the strength of the relationship and whether it is positive or negative. In addition to the degree of correlation between two sets of information, the geographer often needs some means of generalizing a relationship in order to measure the rate at which the two variables are changing together; this enables him to estimate an unknown value of one variable from a known value of the other. It is necessary, therefore, to fit lines to the belts of points making up a bivariate distribution. When this has been done, unknown values of x and y can be estimated by interpolation. Extrapolation or prediction beyond the range of values of known points on

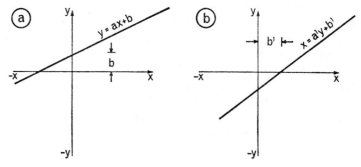

Figure 7.5 The derivation of the equations for straight lines

a scatter diagram is not advisable unless it can be assumed safely that the factors governing the particular relationship do not alter with more extreme values of the variables. Of course, estimates of one variable from the other are unnecessary with cases like the Rousay farms as the information we have is not a sample but a complete population; there are no unknown units. Nevertheless, we may wish to generalise the relationship in terms of a best fitting line.

A line which best fits a scatter of points is known as a *regression line* and may be calculated by the *method of least squares*, explained below. The equation of a straight line may be written as $y = ax + b$, where a and b are constants, a being the slope of the line (which in regression analysis is known as the *regression coefficient*) and b the intercept of the line on the y axis (Figure 7.5(a)). A straight line also may be represented by the equation $x = a'y + b'$, where a' is the gradient of the line and b' the intercept on the x axis (Figure 7.5(b)). The intercepts, of course, may be negative.

The line which best fits a belt of points may not be straight, but curved. Calculation of the equations of such curvilinear regression lines is beyond the scope of an introductory text. If the points on a scatter diagram approximate to a straight line, the best fitting straight line can be calculated to some advantage, although it must be borne in mind that the fit is only approximate. For any bivariate distribution, there are in fact two lines of best fit, dependent upon the method by which 'best fit' is defined. If the variables are denoted by x and y, the lines are known as the regression of y on x, which is used for estimating y for given values of x, and the regression of x on y, used to estimate x for given values of y. The regression of y on x, which has the form $y = ax + b$, is the line for which the sum of the squares of the *residuals* of y for given values of x is at a minimum (hence the term *least squares method*). A residual is the distance between a point and the line, measured at right angles to one of the axes, in this case the x axis (Figure 7.6(a)). In the regression of x on y, which has the form $x = a'y + b'$, the sum of the squares of the residuals of x for given values of y is at a minimum, and the residuals are perpendicular to the y axis (Figure 7.(6b)).

For the regression of y on x, the values of a and b can be calculated from two simultaneous equations, whose formal derivation need not concern us here. These '*normal equations*' have the form

$$\Sigma y = a \Sigma x + nb$$
$$\Sigma xy = a \Sigma x^2 + b \Sigma x$$

The two unknown values, a and b, therefore are calculated easily. a' and b' can be calculated in the same way from similar normal equations for the regression of x on y.

By dividing through by n, the first normal equation can be re-written as $\dfrac{\Sigma y}{n} = a \dfrac{\Sigma x}{n} + b$, which in turn can be written as $\bar{y} = a\bar{x} + b$. The point $(\bar{x}, \bar{y})$, known as the *mean of the array of points*, therefore satisfies the equation $y = ax + b$ and the regression of y on x passes through the mean of the array. It can be demonstrated in a similar manner that the regression of x on y passes through $(\bar{x}, \bar{y})$, so that the regression lines cross at the mean of the array. The angle between two regression lines is a measure of the strength of correlation. If there is a functional relationship, there will be no residuals and the lines will coincide. On the other hand, if the regression lines are at right angles, a complete absence of correlation is indicated.

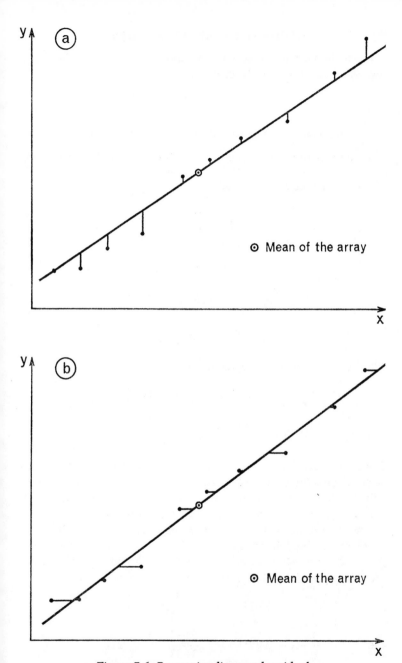

Figure 7.6 Regression lines and residuals
(a) *Residuals of the regression of y on x*
(b) *Residuals of the regression of x on y*

From the normal equations a general value of the coefficient of regression of y on x can be calculated:

$$a = \frac{\sigma_{xy}}{\sigma_x{}^2}$$

As the regression line passes through the mean of the array, its equation may be written as $(y - \bar{y}) = \dfrac{\sigma_{xy}}{\sigma_x{}^2}(x - \bar{x})$ and the regression of x on y may be expressed as

$$(x - \bar{x}) = \frac{\sigma_{xy}}{\sigma_y{}^2}(y - \bar{y}).$$

But as $R_{xy} = \dfrac{\sigma_{xy}}{\sigma_x \sigma_y}$,

$\dfrac{\sigma_{xy}}{\sigma_x{}^2}$ may be written as $R_{xy}\dfrac{\sigma_y}{\sigma_x}$

and $\dfrac{\sigma_{xy}}{\sigma_y{}^2}$ as $R_{xy}\dfrac{\sigma_x}{\sigma_y}$

Alternative forms of the regression equations thus are

$$(y - \bar{y}) = R_{xy}\frac{\sigma_y}{\sigma_x}(x - \bar{x})$$

and

$$(x - \bar{x}) = R_{xy}\frac{\sigma_x}{\sigma_y}(y - \bar{y})$$

The constants in the equations can be calculated rapidly if R_{xy} has been calculated beforehand, as σ_x and σ_y also will have been determined. These alternative equations avoid the solution of two sets of normal equations in determination of the constants a, b, a' and b'.

The relationship between arable acreage and cattle numbers can now be generalised in terms of regression lines. With the information $\bar{y} = 37 \cdot 25$, $\bar{x} = 54 \cdot 25$, $\sigma_y = 35 \cdot 58$, $\sigma_x = 52 \cdot 34$ and $R_{xy} = 0 \cdot 93$, the regression equations can be written as

$$y - 37 \cdot 25 = 0 \cdot 93 . \frac{35 \cdot 58}{52 \cdot 34}(x - 54 \cdot 25)$$

and

$$x - 54 \cdot 25 = 0 \cdot 93 . \frac{52 \cdot 34}{35 \cdot 58}(y - 37 \cdot 25)$$

or

$$y = 0 \cdot 6335x + 2 \cdot 881$$

and

$$x = 1 \cdot 3710y + 3 \cdot 171$$

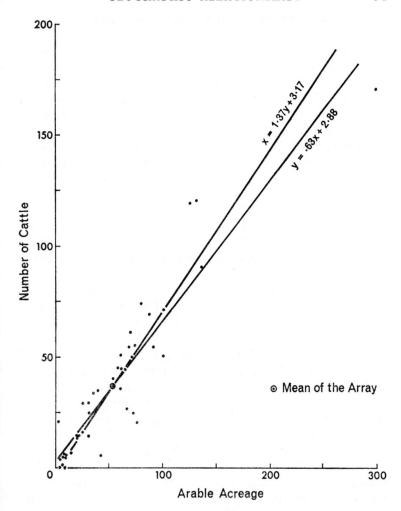

Figure 7.7 Regression lines for the arable acreage and cattle numbers of Rousay farms

These two lines are shown in Figure 7.7; it can be seen that, with such a high coefficient of correlation, they are very close together. They generalise the relationship between number of cattle and arable acreage on Rousay holdings under the conditions prevailing at the time of data collection. They show that the rate of increase in cattle numbers is greater than that of arable acreage, i.e. a farm with twice as much arable land as another has more than twice the number of cattle.

In general, should we wish to estimate a value of y from a value of x then we would use the regression of y on x; this indicates how the number of cattle (y) depends on the arable acreage (x). In this case x is the assumed independent variable and y the dependent. However, this does not imply that a change in x *causes* a change in y; the value of y is merely being estimated from a known or assumed value of x. In other words, the residuals are being taken as the errors in y, since it is assumed that the values of x are known accurately. Should we wish to estimate the value of x from y, we would use the regression of x on y; y would be the independent and x the dependent variable. In the case of Rousay farms, there are no unknown values of x and y to estimate, but we could ask the hypothetical questions: under these conditions, how many cows could a given arable acreage support and how much arable land is needed to support a given number of cows?

It must be stressed that the value of the independent variable derived from the graph for a known value of the assumed independent variable is an *estimate*, as the line itself is a best-fit or best estimate. As was the case with other estimates it is desirable to know the standard error of the estimate. This is equal to $\sigma\sqrt{1 - R_{xy}^2}$, where σ is the standard deviation of the unknown variable and R_{xy} the coefficient of correlation. The properties of standard errors were discussed in Chapter 4. There is a 95 per cent probability that the actual values of cattle number and arable acreage will depart from their regression lines by not more than two standard errors. As the standard error is the same for all estimates in a given situation it (or multiples of it) can be represented on the regression graph as lines parallel to the regression line, known as *confidence limits*. Figure 7.8 shows the confidence limits for the regression of cattle numbers on arable acreage. The graph indicates that, given a 95 per cent probability level, there would be an estimated 66 ± 26 cattle for 100 acres of arable land.

Geographical phenomena may vary with distance or time, and correlation and regression techniques can usefully be employed in their analysis, the former to establish whether significant trends exist and the latter to establish the magnitude of any such trends. In this method of analysis there is really only one variable, which by convention is represented on the y axis, the x axis being devoted to distance or time. We have information about the number of sheep in the parish of Rousay for the period 1870–1966 (Appendix A). (It should be noted that the parish is larger than the island of Rousay, since it consists of several islands in addition to Rousay.) In such

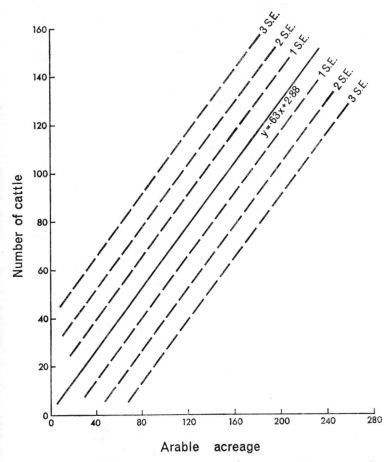

Figure 7.8 The regression of cattle numbers on arable acreage for Rousay farms, and its confidence limits

cases, as with considerations of distance, it is desirable to establish significant trends. If the points for each year are joined together, a rather irregular upward trend is revealed (Fig. 7.9). The simplest way to generalise and simplify this is to use the technique of *running means* (also known as *moving means*), which is a method of reducing the irregularities.

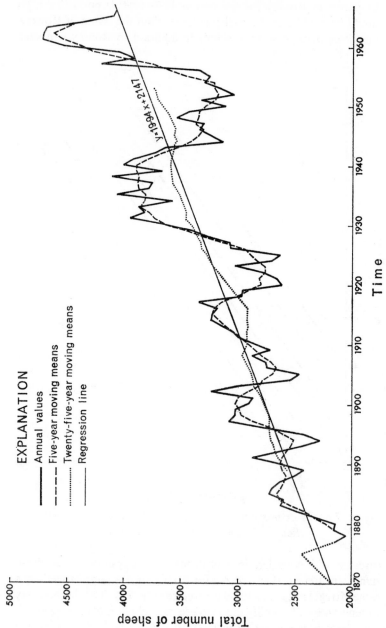

Figure 7.9 The number of sheep in Rousay parish, 1870-1966

The mean number of sheep for the period 1877–81 may be determined by adding together the values for each year and dividing the total by 5. A similar procedure will reveal the mean value for the period 1878–82. The difference between the means of the two five-year periods clearly will reflect the different numbers of sheep existing in 1877 and 1882, since the values for the intervening years are included in both means. Continuing the process, a series of means may be calculated for 'overlapping' five-year periods until the final one 1962–6. A graph of the running means, the values for the overlapping periods, shows the same upward trend as that of the individual annual values, but the most striking fluctuations are smoothed out; the 'crests' and 'troughs' of the annual graph are flattened.

The longer the periods used for the running mean values, the larger are the fluctuations of single years which can be smoothed out. A plot of running mean annual values of sheep for overlapping ten-year periods would be even smoother than that for the five-year periods shown in Figure 7.9. Instead of making an individual appearance in the graph, exceptionally high or low values in particular years are 'averaged out' several times in running means, so that 'exceptions to the rule' are less likely to obscure specific upward or downward trends than if they stood alone. Running mean plots thus show generalised trends. The choice of period to be used for the overlapping intervals is governed by the level of approximation which is required or which is judged acceptable. The twenty-five-year running mean line for Rousay sheep virtually eliminates all irregularities (Figure 7.9).

Although revealing apparent changes or trends, a plot of running means does not indicate their significance; nor does it suggest whether or not they may be the result of chance. We cannot use the running means of Figure 7.9 to assess the probability that any of the apparent changes are within a certain range of chance. We must return to the original annual data to obtain the best fit rate of change, using for the purpose the regression of y (sheep) on x (time). The significance of the relationship will be indicated in the usual way by the value of the correlation coefficient, R_{xy}. Naturally, we may only fit a regression line to data which justifies the assumption of a linear trend (rate of change). The regression line for numbers of sheep (y) on the period in years 1868–1966 (x) has the equation $y = 19 \cdot 94x + 2147$, where $1870 = 1$, $1880 = 11$, etc. The value of the correlation coefficient is $R_{xy} = 0.8385$. The relationship indicated by the regression of y on x thus is likely to be significant, i.e. the assumption, implicit in plotting them against each other, that sheep numbers correlate with time, is supported. Time itself is not the cause of the change, but the recognisable trend illustrates a close relationship between the increase in numbers of sheep and the passage of time.

In a time series such as this, the nature of the data available may mask some recurrent trends. For instance, if sheep numbers vary seasonally, data collected annually on one date will not reveal the fact. Annual returns may mask a recurring cycle of increasing numbers in summer and decreasing numbers in winter. Care must be taken, therefore, to consider the nature of the time interval between successive presentations of data used in plotting running means.

In this chapter, we have concentrated attention on cases involving two variables only and on relationships which can be represented by a straight line. Often, however, several variables may be interrelated with each other, so that problems of multiple correlation and regression arise. Relationships may be curvilinear rather than linear. Although the procedure for analysing multiple linear regression (variations in y being related to several x variates) is analogous to that for simple linear regression, and analysis of curvilinear relationships involves no very different concepts from those used in considering linear relationships, the procedures lead to more complex arithmetic. Calculations frequently cannot be handled with simple equipment and may be beyond the scope of a desk calculator. For this reason, multiple and curvilinear relationships will not be considered further.

8. Reflections

In this book, we have dealt with the ordering and summarising of geographical data. This can take us only part of the way in a geographical investigation; the final stage is the explanation, or the attempt at explanation, of why the phenomena we study are as they are. Techniques are not an end in themselves, although some of them take a project further towards the stage of explanation than do others. The methods of analysis outlined in the preceding chapters can be applied to a very wide range of geographical data. The results of their application to the three sets of data raise many points, clarify some situations and provide a few answers.

The island of Rousay is a small, clearly defined area about which we know a great deal. The basic geographical units are holdings, which have various attributes (acreage, sheep, cattle, etc.). Definite statements may be made about the holdings, concerning, for instance, the number of cattle on them, or the acreage of land in arable use. The available information about Rousay, like all geographical information, can be classified into three groups—aspects of size, location and time.

The size of Rousay holdings is better summarised by the median than by the mean as the distribution is markedly skewed. Why does the size distribution of the holdings have this form? In seeking an answer to this question, our personal judgement must form the basis for further investigation. We must set up a hypothesis for consideration. We may propose that the pattern of farm sizes is the result of particular historical events or of certain social trends. Testing these ideas requires much more information than has been presented in this text. A similar situation exists with regard to the discussion of the spatial distribution of the holdings; once the precise distribution has been described, the real questions emerge. Why are houses and holdings clustered into groups? We might suggest that physical factors influence the settlement pattern. This hypothesis then would have to be tested by comparing the distribution of physical factors, especially those restrictive of settlement or agriculture, with that of the farms and other houses. Again, we might suggest that topography, drainage or even historical circumstances have been responsible for the peripheral location of settlement on the island, a hypothesis which would necessitate the collection of further data for testing. The temporal data, such as the progressive increase of the number of sheep, are no exception in posing the question 'why?'. Once more, geographical knowledge and judgement become important in the formulation of hypotheses.

The explanation of particular distributions in isolation is rarely possible, as all elements in a situation invariably are related to a greater or lesser degree. In the case of Rousay, the most obvious way to tackle the problem of association is in terms of the agricultural (economic) geography of the island. This appears to be primarily pastoral with an arable element. In testing all the variables for correlation, we found a numerical association between numbers of cattle and arable acreage. By itself, this explains nothing. Does it have any real meaning? Further studies would be necessary before this question could be answered in a satisfactory manner. It might be essential to use a questionnaire in order to obtain more information about each holding. We must always bear in mind the possibility, however remote, that an association results from chance. Wherever possible, the results of one study should be checked by another line of approach. Geographical experience and interpretation cannot be abandoned because of one correlation.

The Manchester clothing industry provides the basis for a larger-scale study than that of Rousay. The fundamental unit here is the firm, about which very little is known in the first instance—merely its location. Size attributes, such as the number of workers, have to be established by careful sampling, so that limits of accuracy are imposed even at the stage of descriptive presentation of data. As always, geographical conclusions based on sample information must be approached with extreme caution. Because the locations of all firms were known, sampling was not necessary for calculation of mean centres and standard distances. However, the results raise questions rather than supplying answers. Why are firms located in particular places? Why, for instance, are the standard distances of shirt- and dress-manufacturing firms larger than those of other branches of the clothing industry? Why is the mean centre of dress manufacture offset to the east of the city centre, away from the other mean centres?

It should not be thought that geographical studies consist solely of summarising the available data and hoping that problems emerge. Although this might be possible in some extremely limited situations, it is hardly possible where enormous amounts of data exist. In studies such as that of the clothing industry, subjective judgement must enter at an early stage, in the selection of the data to be considered and the techniques to be applied in their analysis or interpretation. Before we could attempt to establish whether significant differences exist between firms near the city centre and those further away, we had to define the limits of the inner and outer zones. This was a matter of judgement as no clear line separates inner firms from outer ones. To some degree, therefore, the square enclosing the inner firms is arbitrary (Figure 4.3); in a limited study, not all possible dividing lines can be considered.

Studies revealed that the numbers of employees of firms within the inner clothing area of the Manchester conurbation differed from those of firms in the outer zone, but the difference was not statistically significant. In this respect, therefore, the areas have not been shown to be different. The method of analysis used helps us to avoid drawing a false conclusion, rather than to arrive at a valid one. On the other hand, at the probability level which we judged acceptable (1 per cent), we found the two areas to be different in terms of the proportions of firms engaged in various branches of the clothing industry. The crucial question is why the proportions differ. Again, it is the location of individual firms which must be studied. A variety of methods must be used in studying the location of firms within the conurbation and the dominance of the conurbation in the clothing industry of the country, demonstrated by the value of the location quotient.

We have presented a very limited approach to a potentially large study. Many aspects of the Manchester clothing industry, such as the mobility of firms and of workers, the output of firms, and the considerations upon which location is decided, have been left untouched. However, a simple quantified base is essential, especially when the amount of available information could be very large. This must be borne in mind constantly as the 'information explosion' in geography continues in the future.

Census returns, dating from 1831, are packed with information, from numbers of households to employment data, all of it inviting analysis. Our consideration of Census material relating to the population of Lancashire has been extremely limited; a geographical study which makes use only of the present distribution of population by administrative areas can reveal little. The essential point relating to this particular study is that, unlike farms or firms, administrative areas pose difficulties of comparability. The geographer frequently finds that data are aggregated in irregular anachronistic units of unequal size. The definition of precise areas of high or low population density from Census material is virtually impossible, even when a County Borough, such as Salford, is broken down into wards. Nevertheless, the Census returns are a rich store of material for analysis by the social geographer, the applied geographer, the planner and other workers. Population changes resulting from migration, or from slum clearance and the building of new towns in previously rural areas, may be studied with the aid of successive Census reports. Overcrowded or other socially depressed areas may be defined, forming the basis for planning decisions.

All geographical studies involve some form of data, most of which can be expressed in numerical terms. Since conclusions should be drawn only from reliable data, an essential first step in dealing with any problem is to present the data in a reasoned manner and to make a partial analysis, so that inherent errors, limitations and difficulties may be pointed out. Precision in subsequent calculations is worthless if the basic data are unreliable. At all stages, margins of error must be stated. This first step in dealing with the problem may not only reveal some questions to which answers must be sought but also enable us to put into precise terms those questions which were apparent at the beginning of the exercise.

The scale of many geographical problems is such that sampling is essential. It must be carried out with great care. Case studies merely assumed to be 'representative' are not sufficient; samples must be drawn properly. Precision, even if only in assessing limits of confidence, is of paramount importance. Estimates of parameters derived from samples must not be treated as definitive values. The future of geographical studies depends on rigour of analysis. However, this in no way alters what geography is about; the essential geographical problems of why phenomena are arranged as they are, and what relationships they bear to one another, remain unchanged. All that is happening is that techniques of data manipulation are being used more often. Although many of the conclusions reached with the aid of numerical techniques are the same as those reached without them, others are not; those which are confirmed are established more firmly and more precisely than before. The possibilities for geographical studies which are being opened up by the increasing use of statistical techniques make the study of geography in the future an exciting prospect.

Appendix A

*Farm sizes, livestock and arable acreage on holdings
in Rousay, 1966*

Farm No.	Total acreage	Arable acreage	Number of Cattle	Number of Sheep
1	4	4	3	0
2	4	4	0	0
3	4	4	3	0
4	5	5	1	0
5	6	6	4	0
6	8	8	4	0
7	13	8	6	11
8	14	7	6	0
9	20	7	0	14
10	25	21	13	0
11	40	20	12	73
12	40	30	14	0
13	40	25	27	0
14	44	2	21	0
15	50	35	33	15
16	51	31	24	8
17	59	19	14	31
18	66	54	40	64
19	70	30	27	0
20	72	61	51	0
21	73	24	16	0
22	76	65	44	0
23	100	65	27	0
24	107	72	50	0
25	110	75	20	71
26	114	42	6	25
27	120	67	53	94
28	150	14	7	27
29	158	70	61	96
30	160	60	36	45
31	171	71	24	14
32	187	73	55	0
33	198	58	45	66
34	212	62	44	148
35	219	39	35	30
36	230	135	90	85
37	310	100	71	122
38	386	86	68	207
39	391	91	53	81
40	435	80	73	80
41	600	100	50	30

Farm No.	Total acreage	Arable acreage	Number of Cattle	Sheep
42	960	125	118	212
43	1900	132	120	100
44	3000	300	170	617

Total number of sheep in Rousay parish, 1870–1966 (from June 4th returns)

1870	2177	1895	2430	1919	2970	1943	3658
1871	—	1896	2788	1920	2630	1944	3156
1872	—	1897	3084	1921	2678	1945	3244
1873	—	1898	3039	1922	2778	1946	3343
1874	—	1899	3039	1923	3045	1947	3315
1875	2431	1900	2918	1924	2719	1948	3565
1876	—	1901	2890	1925	2651	1949	3434
1877	2156	1902	3248	1926	3077	1950	3113
1878	2036	1903	2970	1927	3090	1951	3338
1879	2154	1904	2555	1928	3295	1952	3047
1880	2135	1905	2482	1929	3410	1953	3142
1881	2298	1906	2746	1930	3656	1954	3313
1882	2478	1907	2768	1931	3948	1955	3276
1883	2625	1908	2877	1932	3824	1956	3343
1884	2607	1909	2744	1933	3887	1957	4228
1885	2742	1910	2863	1934	3593	1958	3916
1886	2704	1911	3003	1935	4078	1959	4042
1887	2635	1912	3053	1936	3791	1960	4542
1888	2501	1913	3141	1937	3769	1961	4720
1889	2418	1914	3252	1938	4111	1962	4718
1890	2638	1915	3244	1939	3673	1963	4747
1891	2888	1916	3200	1940	4026	1964	4373
1892	2676	1917	3345	1941	3921	1965	4103
1893	2438	1918	2990	1942	3739	1966	4090
1894	2298						

Appendix B

The Manchester Clothing Industry, 1966

Total number of firms	826
Number of clothing workers in Britain (c_1)	365 700
Number of clothing workers in N.W. England (c_2)	73 600
Number of clothing workers in the Manchester Conurbation (c_3)	43 514
Number of workers in Britain (n_1)	24 380 000
Number of workers in N.W. England (n_2)	3 228 000
Number of workers in the Manchester Conurbation (n_3)	1 231 250

Appendix C

Population and areas of County Boroughs, Municipal Boroughs, Urban Districts and Rural Districts in Lancashire, 1961

County Boroughs	Population	Acreage
Liverpool	745 750	27 810
Manchester	661 791	27 255
Bolton	160 789	15 280
Salford	155 090	5 203
Blackpool	153 185	8 609
Oldham	115 346	6 392
Preston	113 341	6 357
St. Helens	108 674	8 865
Blackburn	106 242	8 088
Rochdale	85 787	9 556
Bootle	82 773	3 057
Southport	82 004	9 652
Burnley	80 559	4 686
Wigan	78 690	5 083
Warrington	75 964	4 520
Barrow-in-Furness	64 927	11 002
Bury	60 149	7 433
TOTAL	2 931 061	168 848

Municipal Boroughs and Urban Districts	Population	Acreage
Huyton-with-Roby	63 089	3 055
Stretford M.B.	60 364	3 533
Crosby M.B.	59 166	4 785
Middleton M.B.	56 688	5 172
Widnes	52 186	5 746
Kirkby	52 088	4 672
Ashton-under-Lyne M.B.	50 154	4 135
Lancaster M.B.	48 235	4 873
Leigh M.B.	46 174	6 359
Eccles M.B.	43 173	3 417
Urmston	43 068	4 799
Swinton and Pendlebury M.B.	40 470	3 362
Worsley	40 393	7 240
Morecambe and Heysham M.B.	40 228	3 794
Accrington M.B.	39 018	4 418
Lytham St. Anne's M.B.	36 189	5 814
Prestwich M.B.	34 209	2 421
Chadderton	32 568	3 014

Municipal Boroughs and Urban Districts	Population	Acreage
Nelson M.B.	32 292	3 445
Chorley M.B.	31 315	4 283
Denton	31 089	2 593
Darwen M.B.	29 475	5 959
Fleetwood M.B.	27 686	2 565
Farnworth M.B.	27 502	1 679
Radcliffe M.B.	26 726	4 957
Droylsden	25 461	1 245
Litherland	24 871	1 210
Heywood M.B.	24 090	8 508
Rawtenstall M.B.	23 890	9 528
Ormskirk	21 828	15 608
Newton-le-Willows	21 768	3 105
Golborne	21 310	7 567
Thornton Cleveleys	20 648	3 358
Failsworth	19 819	1 679
Atherton	19 756	2 265
Colne	19 430	5 939
Leyland	19 413	3 804
Hindley	19 396	2 610
Ashton-in-Makerfield	19 262	6 266
Walton-le-Dale	18 964	4 733
Ince-in-Makerfield	18 019	2 321
Bacup M.B.	17 308	6 121
Tyldesley	16 813	5 175
Westhoughton	16 260	5 560
Horwich	16 078	3 257
Fulwood	16 016	3 164
Irlam	15 371	4 717
Royton	14 474	2 148
Whitefield	14 372	3 391
Haslingden M.B.	14 360	8 203
Ramsbottom	13 817	9 526
Turton	13 698	17 334
Prescot	13 079	871
Poulton-le-Fylde	12 726	2 272
Crompton	12 708	2 865
Clitheroe M.B.	12 158	2 386
Audenshaw	12 122	1 241
Haydock	12 074	2 395
Oswaldtwistle	11 918	4 885
Formby	11 734	5 613
Great Harwood	10 718	2 868
Orrell	10 664	1 616
Littleborough	10 552	7 855
Ulverston	10 527	3 206
Dalton-in-Furness	10 316	8 022
Kearsley	10 296	1 727
Padiham	9 899	975
Mossley M.B.	9 776	3 661
Standish-with-Langtree	9 692	3 266
Milnrow	8 129	5 194

Municipal Boroughs and Urban Districts	Population	Acreage
Upholland	7 452	4 684
Whitworth	7 064	4 483
Brierfield	7 018	807
Billinge-and-Winstanley	6 945	4 596
Aspull	6 748	1 905
Clayton-le-Moors	6 421	1 060
Skelmersdale	6 309	1 941
Abram	6 004	1 979
Church	5 888	528
Tottington	5 649	2 542
Rishton	5 433	2 879
Rainford	5 385	5 877
Little Lever	5 085	807
Kirkham	4 819	939
Longridge	4 686	3 285
Barrowford	4 644	1 387
Wardle	4 608	3 192
Adlington	4 276	1 062
Carnforth	4 113	1 504
Lees	3 730	288
Blackrod	3 606	2 392
Grange	3 125	1 883
Withnell	2 849	4 186
Preesall	2 357	3 277
Trawden	1 952	6 815
TOTAL	1 875 271	379 510

Rural Districts	Population	Acreage
West Lancashire	55 763	65 620
Whiston	43 786	23 786
Preston	43 592	49 754
Warrington	30 732	22 350
Chorley	28 567	41 117
Fylde	17 370	33 264
North Lonsdale	16 598	127 448
Burnley	16 035	39 849
Blackburn	15 053	19 469
Garstang	14 390	57 491
Lancaster	14 018	53 212
Wigan	10 157	11 695
Clitheroe	8 799	32 170
Lunesdale	8 224	76 267
TOTAL	323 084	653 492
GRAND TOTAL	5 129 416	1 201 850

Appendix D

The normal curve

The equation of the normal curve, relating the distance (y) on the vertical axis at any point (x) on the horizontal axis to that point is:

$$y = \frac{1}{\sigma\sqrt{2\pi}}\, e^{-\frac{(x-\bar{x})^2}{2\sigma^2}}$$

where σ is the standard deviation and e and π are mathematical constants. Thus, y is a function of x, its mean and its standard deviation, and the form of normal curves always is the same bell shape. However, normal curves differ from each other in terms of $\bar{x}$ and σ. A *standard form* of the equation for the normal distribution eliminates these two parameters which cause variability and enables maximum use to be made of the curve. A standard normal distribution is defined as a normal distribution which has $\bar{x} = 0$ and $\sigma = 1$. Its equation then is

$$y = \frac{1}{\sqrt{2\pi}}e^{-\frac{z^2}{2}}$$

where the symbol $z = \dfrac{(x-\bar{x})}{\sigma}$. All values of deviation from the mean then are measured in terms of the standard deviation of the distribution, i.e. each individual deviation is expressed as a proportion of the standard deviation. The area under the curve between any two values of z represents the probability that an item chosen at random from the data will fall between the two values of the variable (x) which correspond to these two values of z. The tables on p. 108 facilitate calculation of the area under the curve below any given value of z, and the value of y for a given value of z.

The ordinate, Y of the normal probability function

Z	·00	·02	·04	·06	·08
·0	·3989	·3989	·3986	·3982	·3977
·1	·3970	·3961	·3951	·3939	·3925
·2	·3910	·3894	·3876	·3857	·3836
·3	·3814	·3790	·3765	·3739	·3712
·4	·3683	·3653	·3621	·3589	·3555
·5	·3521	·3485	·3448	·3410	·3372
·6	·3332	·3292	·3251	·3209	·3166
·7	·3123	·3079	·3034	·2989	·2943
·8	·2897	·2850	·2803	·2756	·2709
·9	·2661	·2613	·2565	·2516	·2468
1·0	·2420	·2371	·2323	·2275	·2227
1·1	·2179	·2131	·2083	·2036	·1989
1·2	·1942	·1895	·1849	·1804	·1758
1·3	·1714	·1669	·1626	·1582	·1539
1·4	·1497	·1456	·1415	·1374	·1334
1·5	·1295	·1257	·1219	·1182	·1145
1·6	·1109	·1074	·1040	·1006	·0973
1·7	·0940	·0909	·0878	·0848	·0818
1·8	·0790	·0761	·0734	·0707	·0681
1·9	·0656	·0632	·0608	·0584	·0562
2·0	·0540	·0519	·0498	·0478	·0459
2·1	·0440	·0422	·0404	·0387	·0371
2·2	·0355	·0339	·0325	·0310	·0297
2·3	·0283	·0270	·0258	·0246	·0235
2·4	·0224	·0213	·0203	·0194	·0184
2·5	·0175	·0167	·0158	·0151	·0143
2·6	·0136	·0129	·0122	·0116	·0110
2·7	·0104	·0099	·0093	·0088	·0084
2·8	·0079	·0075	·0071	·0067	·0063
2·9	·0060	·0056	·0053	·0050	·0047

Linear interpolation sufficient

$$Y = \frac{1}{\sqrt{2\pi}} e^{-\frac{1}{2}Z^2}$$

Z is the standardized variable with zero mean and unit standard deviation

For negative values of Z note: $P(Z) = 1 - P(-Z)$

The integral, P, of the normal probability function

Z	·00	·01	·02	·03	·04	·05	·06	·07	·08	·09
·0	·5000	·5040	·5080	·5120	·5100	·5199	·5239	·5279	·5319	·5359
·1	·5398	·5438	·5478	·5517	·5557	·5596	·5636	·5675	·5714	·5753
·2	·5793	·5832	·5871	·5910	·5948	·5987	·6026	·6064	·6103	·6141
·3	·6179	·6217	·6255	·6293	·6331	·6368	·6406	·6443	·6480	·6517
·4	·6554	·6591	·6628	·6664	·6700	·6736	·6772	·6808	·6844	·6879
·5	·6915	·6950	·6985	·7019	·7054	·7088	·7123	·7157	·7190	·7224
·6	·7267	·7291	·7324	·7357	·7389	·7422	·7454	·7486	·7517	·7549
·7	·7580	·7611	·7642	·7673	·7704	·7734	·7764	·7794	·7823	·7852
·8	·7881	·7910	·7939	·7967	·7995	·8023	·8051	·8078	·8106	·8133
·9	·8159	·8186	·8212	·8238	·8264	·8289	·8315	·8340	·8365	·8389
1·0	·8413	·8438	·8461	·8485	·8508	·8531	·8554	·8577	·8599	·8621
1·1	·8643	·8665	·8686	·8708	·8729	·8749	·8770	·8790	·8810	·8830
1·2	·8849	·8869	·8888	·8907	·8925	·8944	·8962	·8980	·8997	·9015
1·3	·9032	·9049	·9066	·9082	·9099	·9115	·9131	·9147	·9162	·9177
1·4	·9192	·9207	·9222	·9236	·9251	·9265	·9279	·9292	·9306	·9319
1·5	·9332	·9345	·9357	·9370	·9382	·9394	·9406	·9418	·9429	·9441
1·6	·9452	·9463	·9474	·9484	·9495	·9505	·9515	·9525	·9535	·9545
1·7	·9554	·9564	·9573	·9582	·9591	·9599	·9608	·9616	·9625	·9633
1·8	·9641	·9649	·9656	·9664	·9671	·9678	·9686	·9693	·9699	·9706
1·9	·9713	·9719	·9726	·9732	·9738	·9744	·9750	·9756	·9761	·9767
2·0	·9772	·9778	·9783	·9788	·9793	·9798	·9803	·9808	·9812	·9817
2·1	·9821	·9826	·9830	·9834	·9838	·9842	·9846	·9850	·9854	·9857
2·2	·9861	·9864	·9868	·9871	·9875	·9878	·9881	·9884	·9887	·9890
2·3	·9893	·9896	·9898	·99010	·99036	·99061	·99086	·99111	·99134	·99158
2·4	·99180	·99202	·99224	·99245	·99266	·99286	·99305	·99324	·99343	·99361
2·5	·99379	·99396	·99413	·99430	·99446	·99461	·99477	·99492	·99506	·99520
2·6	·99534	·99547	·99560	·99573	·99585	·99598	·99609	·99621	·99632	·99643
2·7	·99653	·99664	·99674	·99683	·99693	·99702	·99711	·99720	·99728	·99736
2·8	·99744	·99752	·99760	·99767	·99774	·99781	·99788	·99795	·99801	·99807
2·9	·99813	·99819	·99825	·99831	·99836	·99841	·99846	·99851	·99856	·99861

Linear interpolation sufficient

Z	3·0	3·1	3·2	3·3	3·4	3·5	3·6	3·7	3·8	3·9
P	·99865	·99903	·99931	·99952	·99966	·99977	·99984	·99989	·99993	·99995
Y	·00443	·00327	·00238	·00172	·00123	·00087	·00061	·00042	·00029	·00020

Y (ordinate)

Reproduced with permission from Biometrika Tables for Statisticians

Appendix E

Random numbers (*See over page*)

The Use of the Random Numbers Table

The first number to be used, and the direction in which the table is to be read, must be selected by chance. A pin may be dropped onto the table and the number nearest to the point taken as the first one. If the point falls in the top left quarter of the table, read subsequent numbers to the left; if the point falls in the bottom right quarter, read to the right. Read upwards if the point falls in the top right quarter of the table and downwards if it falls in the bottom left quarter.

Assign a number to each item in the parent population, using single numbers, including 0, when the population is 10 or less. Use pairs of numbers (01, 02, etc.) for a population of from 11 to 100 (00 is the 100th member of the group). Assign three numbers to each member if the total population is between 101 and 1000. In drawing a sample, ignore any number already selected and any number larger than the highest assigned value, e.g. with a population of 826, ignore numbers from 827 upwards.

In order to minimise the number of discards when the population size is between 101 and 200, the first number in each triplet may be changed to 0 if it is even and to 1 if it is odd, e.g. with a population of 200, the first line of the table becomes 124 165 153 042 194 etc.

98	17	41	06	54	89	41	62	30	57	47	63	58	14	45	04	10	94	10	68
47	16	63	99	44	65	00	68	81	54	96	62	80	64	28	82	48	21	37	51
42	06	97	97	66	47	53	40	07	61	56	20	27	25	42	24	03	76	93	03
74	39	92	77	79	50	61	46	40	76	61	01	06	84	34	45	74	70	46	78
22	56	75	08	42	01	03	41	71	79	36	28	46	68	25	95	68	53	11	57
50	11	85	02	64	36	32	79	97	24	10	54	11	41	96	81	50	40	14	84
42	57	96	81	80	25	35	76	99	87	74	08	49	41	68	38	61	96	78	03
86	10	86	33	74	28	17	16	95	08	84	24	81	79	42	48	99	34	31	56
12	45	01	07	81	43	75	14	72	55	17	10	98	19	07	31	67	51	90	57
95	99	41	98	45	97	36	12	94	18	92	43	96	17	79	34	04	23	06	12
80	72	49	91	76	28	48	46	91	19	63	25	62	18	22	03	44	82	61	93
55	60	27	55	63	79	80	16	40	25	36	07	71	11	28	48	11	97	11	55
41	04	18	01	24	74	30	32	41	79	02	31	76	91	46	58	27	85	09	03
94	04	05	95	85	17	29	94	59	92	93	05	15	77	81	27	32	98	95	77
95	76	41	15	78	15	57	45	74	26	18	36	99	28	99	20	79	56	34	05
08	27	42	62	48	66	60	68	53	28	77	11	24	22	83	96	15	14	16	24
40	49	72	87	72	66	40	05	76	68	80	99	55	04	56	85	55	04	82	60
59	45	33	13	46	59	96	06	33	85	28	08	27	39	81	66	00	34	62	03
42	26	87	72	79	91	88	42	32	73	78	61	76	43	45	30	43	89	21	29
38	29	47	91	85	65	68	85	83	89	68	74	61	06	53	53	02	93	37	67
35	05	62	23	04	91	70	73	79	20	29	08	87	24	19	94	25	58	43	21
65	72	65	61	76	67	00	00	35	31	48	36	05	57	35	75	26	34	43	25
49	16	40	05	90	04	58	24	30	25	41	76	37	62	98	47	30	99	26	46
52	70	72	73	50	02	45	98	05	27	84	89	40	96	74	79	78	93	50	49

09 79 51 45 32   43 14 65 54 82   70 61 12 83 96   72 16 31 18 68
68 01 19 17 70   69 01 85 05 39   09 32 93 55 22   03 02 90 00 57
96 85 53 24 72   27 77 08 19 02   99 67 06 33 33   08 11 04 18 30
12 03 72 22 53   40 67 16 57 50   14 91 41 91 59   00 88 48 86 85
68 37 73 31 29   41 44 76 29 68   40 23 37 02 37   53 38 83 04 33
29 83 13 90 52   75 65 58 52 28   50 10 88 19 76   74 56 72 32 41
05 57 71 11 94   73 92 15 22 55   18 49 10 49 68   85 33 34 58 94
47 95 00 09 70   54 52 43 69 88   29 48 65 32 72   79 93 85 55 96
75 44 62 54 58   38 38 02 08 38   58 63 11 91 78   47 21 00 46 97
05 54 05 05 58   88 69 52 03 19   32 23 18 48 31   58 11 52 75 11
65 04 96 74 39   53 88 11 40 07   45 47 91 83 69   35 73 09 44 72
81 35 15 92 57   07 76 69 19 45   28 59 51 03 42   67 62 49 38 41
46 72 80 15 20   37 67 00 01 68   15 33 98 47 42   03 43 66 88 13
09 48 30 76 53   72 28 37 00 16   91 34 19 92 47   85 91 55 47 89
21 98 65 70 15   77 87 66 67 54   05 31 50 12 69   96 79 59 71 96
77 52 83 77 58   78 28 33 98 27   94 18 19 37 35   44 13 91 91 89
96 78 10 18 15   03 48 30 24 90   00 48 63 10 50   20 36 98 32 27
96 11 08 60 66   33 43 42 74 06   30 48 19 93 71   37 16 72 11 49
65 56 15 02 97   99 39 57 21 21   76 99 42 81 79   55 37 62 22 41
98 51 84 54 39   18 19 79 38 62   11 12 61 49 56   30 14 01 66 08
24 35 33 81 56   09 83 58 87 62   99 62 74 96 92   59 05 24 99 92
04 53 68 05 37   22 77 97 47 54   29 06 68 75 86   63 64 66 89 09
89 79 84 34 35   21 77 44 91 17   53 28 34 49 30   83 67 43 40 44
61 33 96 28 19   37 46 12 08 98   73 35 50 95 22   68 69 04 03 95

Appendix F

2Q	30	20	10	5	2	1	0·2	0·1
1	1·963	3·078	6·314	12·706	31·821	63·657	318·31	636·62
2	1·386	1·886	2·920	4·303	6·965	9·925	22·326	31·598
3	1·250	1·638	2·353	3·182	4·541	5·841	10·213	12·924
4	1·190	1·533	2·132	2·776	3·747	4·604	7·173	8·610
5	1·156	1·476	2·015	2·571	3·365	4·032	5·893	6·869
6	1·134	1·440	1·943	2·447	3·143	3·707	5·208	5·959
7	1·119	1·415	1·895	2·365	2·998	3·499	4·785	5·408
8	1·108	1·397	1·860	2·306	2·896	3·355	4·501	5·041
9	1·100	1·383	1·833	2·262	2·821	3·250	4·297	4·781
10	1·093	1·372	1·812	2·228	2·764	3·169	4·144	4·587
11	1·088	1·363	1·796	2·201	2·718	3·106	4·025	4·437
12	1·083	1·356	1·782	2·179	2·681	3·055	3·930	4·318
13	1·079	1·350	1·771	2·160	2·650	3·012	3·852	4·221
14	1·076	1·345	1·761	2·145	2·624	2·977	3·787	4·140
15	1·074	1·341	1·753	2·131	2·602	2·947	3·733	4·073
16	1·071	1·337	1·746	2·120	2·583	2·921	3·686	4·015
17	1·069	1·333	1·740	2·110	2·567	2·898	3·646	3·965
18	1·067	1·330	1·734	2·101	2·552	2·878	3·610	3·922
19	1·066	1·328	1·729	2·093	2·539	2·861	3·579	3·883
20	1·064	1·325	1·725	2·086	2·528	2·845	3·552	3·850
21	1·063	1·323	1·721	2·080	2·518	2·831	3·527	3·819
22	1·061	1·321	1·717	2·074	2·508	2·819	3·505	3·792
23	1·060	1·319	1·714	2·069	2·500	2·807	3·485	3·767
24	1·059	1·318	1·711	2·064	2·492	2·797	3·467	3·745
25	1·058	1·316	1·708	2·060	2·485	2·787	3·450	3·725
26	1·058	1·315	1·706	2·056	2·479	2·779	3·435	3·707
27	1·057	1·314	1·703	2·052	2·473	2·771	3·421	3·690
28	1·056	1·313	1·701	2·048	2·467	2·763	3·408	3·674
29	1·055	1·311	1·699	2·045	2·462	2·756	3·396	3·659
30	1·055	1·310	1·697	2·042	2·457	2·750	3·385	3·646
40	1·050	1·303	1·684	2·021	2·423	2·704	3·307	3·551
60	1·046	1·296	1·671	2·000	2·390	2·660	3·232	3·460
120	1·041	1·289	1·658	1·980	2·358	2·617	3·160	3·373
∞	1·0364	1·2815	1·6449	1·9600	2·3263	2·5758	3·0902	3·2905

Degrees of Freedom ν

*Reproduced with permission from
Biometrika Tables for Statisticians*

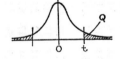

Appendix G

Percentage points of the χ^2 distribution

d.f.	99·5	99·0	97·5	95·0	90·0	75·0	50·0	25·0	10·0	5·0	2·5	1·0	0·5	0·1
1	$3927 \cdot 10^{-8}$	$1571 \cdot 10^{-7}$	$9821 \cdot 10^{-6}$	$3932 \cdot 10^{-6}$	0·01579	0·1015	0·4549	1·323	2·706	3·841	5·024	6·635	7·879	10·828
2	0·01003	0·02010	0·05064	0·1026	0·2107	0·5754	1·386	2·773	4·605	5·991	7·378	9·210	10·597	13·816
3	0·07172	0·1148	0·2158	0·3518	0·5844	1·213	2·366	4·108	6·251	7·815	9·348	11·345	12·838	16·266
4	0·2070	0·2971	0·4844	0·7107	1·064	1·923	3·357	5·385	7·779	9·488	11·143	13·277	14·860	18·467
5	0·4117	0·5543	0·8312	1·145	1·610	2·675	4·351	6·626	9·236	11·070	12·833	15·086	16·750	20·515
6	0·6757	0·8721	1·237	1·635	2·204	3·455	5·348	7·841	10·645	12·592	14·449	16·812	18·548	22·458
7	0·9893	1·239	1·690	2·167	2·833	4·255	6·346	9·037	12·017	14·067	16·013	18·475	20·278	24·322
8	1·344	1·646	2·180	2·733	3·490	5·071	7·344	10·219	13·362	15·507	17·535	20·090	21·955	26·125
9	1·735	2·088	2·700	3·325	4·168	5·899	8·343	11·389	14·684	16·919	19·023	21·666	23·589	27·877
10	2·156	2·558	3·247	3·940	4·865	6·737	9·342	12·549	15·987	18·307	20·483	23·209	25·188	29·588
11	2·603	3·053	3·816	4·575	5·578	7·584	10·341	13·701	17·275	19·675	21·920	24·725	26·757	31·264
12	3·074	3·571	4·404	5·226	6·304	8·438	11·340	14·845	18·549	21·026	23·337	26·217	28·300	32·909
13	3·565	4·107	5·009	5·892	7·041	9·299	12·340	15·984	19·812	22·362	24·736	27·688	29·819	34·528
14	4·075	4·660	5·629	6·571	7·790	10·165	13·339	17·117	21·064	23·685	26·119	29·141	31·319	36·123
15	4·601	5·229	6·262	7·261	8·547	11·036	14·339	18·245	22·307	24·996	27·488	30·578	32·801	37·697
16	5·142	5·812	6·908	7·962	9·312	11·912	15·338	19·369	23·542	26·296	28·845	32·000	34·267	39·252
17	5·697	6·408	7·564	8·672	10·085	12·792	16·338	20·489	24·769	27·587	30·191	33·409	35·718	40·790
18	6·265	7·015	8·231	9·390	10·865	13·675	17·338	21·605	25·989	28·869	31·526	34·805	37·156	42·312
19	6·844	7·633	8·907	10·117	11·651	14·562	18·338	22·718	27·204	30·143	32·852	36·191	38·582	43·820
20	7·434	8·260	9·591	10·851	12·443	15·452	19·337	23·828	28·412	31·410	34·170	37·566	39·997	45·315
21	8·034	8·897	10·283	11·591	13·240	16·344	20·337	24·935	29·615	32·670	35·479	38·932	41·401	46·797
22	8·643	9·542	10·982	12·338	14·041	17·240	21·337	26·039	30·813	33·924	36·781	40·289	42·796	48·268
23	9·260	10·196	11·688	13·091	14·848	18·137	22·337	27·141	32·007	35·172	38·076	41·638	44·181	49·728
24	9·886	10·856	12·401	13·848	15·659	19·037	23·337	28·241	33·196	36·415	39·364	42·980	45·558	51·179
25	10·520	11·524	13·120	14·611	16·473	19·939	24·337	29·339	34·382	37·652	40·646	44·314	46·928	52·620
26	11·160	12·198	13·844	15·379	17·292	20·843	25·336	30·434	35·563	38·885	41·923	45·642	48·290	54·052
27	11·808	12·879	14·573	16·151	18·114	21·749	26·336	31·528	36·741	40·113	43·194	46·963	49·645	55·476
28	12·461	13·565	15·308	16·928	18·939	22·657	27·336	32·620	37·916	41·337	44·461	48·278	50·993	56·892
29	13·121	14·256	16·047	17·708	19·768	23·567	28·336	33·711	39·087	42·557	45·722	49·588	52·336	58·302
30	13·787	14·954	16·791	18·493	20·599	24·478	29·336	34·800	40·256	43·773	46·979	50·892	53·672	59·703
40	20·707	22·164	24·433	26·509	29·050	33·660	39·335	45·616	51·805	55·758	59·342	63·691	66·766	73·402
50	27·991	29·707	32·357	34·764	37·689	42·942	49·335	56·334	63·167	67·505	71·420	76·154	79·490	86·661
60	35·535	37·485	40·482	43·188	46·459	52·294	59·335	66·981	74·397	79·082	83·298	88·379	91·952	99·607
70	43·275	45·442	48·758	51·739	55·329	61·698	69·334	77·577	85·527	90·531	95·023	100·425	104·215	112·317
80	51·172	53·540	57·153	60·391	64·278	71·144	79·334	88·130	96·578	101·879	106·629	112·329	116·321	124·839
90	59·196	61·754	65·647	69·126	73·291	80·625	89·334	98·650	107·565	113·145	118·136	124·116	128·299	137·208
100	67·328	70·065	74·222	77·929	82·358	90·133	99·334	109·141	118·498	124·342	129·561	135·807	140·169	149·449

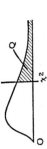

Reproduced with permission from Biometrika Tables for Statisticians

Appendix H

F Distribution: upper 1 per cent points

$d.f. f_2$ \ $d.f. f_1$	1	2	3	4	5	6	7	8	9	10	12	15	20	30	60	∞
1	4052.2	4999.5	5403.3	5624.6	5763.7	5859.0	5928.3	5981.6	6022.5	6055.8	6106.3	6157.3	6208.7	6260.7	6313.0	6366.0
2	98.503	99.000	99.166	99.249	99.299	99.332	99.356	99.374	99.388	99.399	99.416	99.432	99.449	99.466	99.483	99.501
3	34.116	30.817	29.457	28.710	28.237	27.911	27.672	27.489	27.345	27.229	27.052	26.872	26.690	26.505	26.316	26.125
4	21.198	18.000	16.694	15.977	15.522	15.207	14.976	14.799	14.659	14.546	14.374	14.198	14.020	13.838	13.652	13.463
5	16.258	13.274	12.060	11.392	10.967	10.672	10.456	10.289	10.158	10.051	9.8883	9.7222	9.5527	9.3793	9.2020	9.0204
6	13.745	10.925	9.7795	9.1483	8.7459	8.4661	8.2600	8.1016	7.9761	7.8741	7.7183	7.5590	7.3958	7.2285	7.0568	6.8801
7	12.246	9.5466	8.4513	7.8467	7.4604	7.1914	6.9928	6.8401	6.7188	6.6201	6.4691	6.3143	6.1554	5.9921	5.8236	5.6495
8	11.259	8.6491	7.5910	7.0060	6.6318	6.3707	6.1776	6.0289	5.9106	5.8143	5.6668	5.5151	5.3591	5.1981	5.0316	4.8588
9	10.561	8.0215	6.9919	6.4221	6.0569	5.8018	5.6129	5.4671	5.3511	5.2565	5.1114	4.9621	4.8080	4.6486	4.4831	4.3105
10	10.044	7.5594	6.5523	5.9943	5.6363	5.3858	5.2001	5.0567	4.9424	4.8492	4.7059	4.5582	4.4054	4.2469	4.0819	3.9090
11	9.6460	7.2057	6.2167	5.6683	5.3160	5.0692	4.8861	4.7445	4.6315	4.5393	4.3974	4.2509	4.0990	3.9411	3.7761	3.6025
12	9.3302	6.9266	5.9526	5.4119	5.0643	4.8206	4.6395	4.4994	4.3875	4.2961	4.1553	4.0096	3.8584	3.7008	3.5355	3.3608
13	9.0738	6.7010	5.7394	5.2053	4.8616	4.6204	4.4410	4.3021	4.1911	4.1003	3.9603	3.8154	3.6646	3.5070	3.3413	3.1654
14	8.8616	6.5149	5.5639	5.0354	4.6950	4.4558	4.2779	4.1399	4.0297	3.9394	3.8001	3.6557	3.5052	3.3476	3.1813	3.0040
15	8.6831	6.3589	5.4170	4.8932	4.5556	4.3183	4.1415	4.0045	3.8948	3.8049	3.6662	3.5222	3.3719	3.2141	3.0471	2.8684
16	8.5310	6.2262	5.2922	4.7726	4.4374	4.2016	4.0259	3.8896	3.7804	3.6909	3.5527	3.4089	3.2588	3.1007	2.9330	2.7528
17	8.3997	6.1121	5.1850	4.6690	4.3359	4.1015	3.9267	3.7910	3.6822	3.5931	3.4552	3.3117	3.1615	3.0032	2.8348	2.6530
18	8.2854	6.0129	5.0919	4.5790	4.2479	4.0146	3.8406	3.7054	3.5971	3.5082	3.3706	3.2273	3.0771	2.9185	2.7493	2.5660
19	8.1850	5.9259	5.0103	4.5003	4.1708	3.9386	3.7653	3.6305	3.5225	3.4338	3.2965	3.1533	3.0031	2.8442	2.6742	2.4893
20	8.0960	5.8489	4.9382	4.4307	4.1027	3.8714	3.6987	3.5644	3.4567	3.3682	3.2311	3.0880	2.9377	2.7785	2.6077	2.4212
21	8.0166	5.7804	4.8740	4.3688	4.0421	3.8117	3.6396	3.5056	3.3981	3.3098	3.1729	3.0299	2.8796	2.7200	2.5484	2.3603
22	7.9454	5.7190	4.8166	4.3134	3.9880	3.7583	3.5867	3.4530	3.3458	3.2576	3.1209	2.9780	2.8274	2.6675	2.4951	2.3055
23	7.8811	5.6637	4.7649	4.2635	3.9392	3.7102	3.5390	3.4057	3.2986	3.2106	3.0740	2.9311	2.7805	2.6202	2.4471	2.2559
24	7.8229	5.6136	4.7181	4.2184	3.8951	3.6667	3.4959	3.3629	3.2560	3.1681	3.0316	2.8887	2.7380	2.5773	2.4035	2.2107
25	7.7698	5.5680	4.6755	4.1774	3.8550	3.6272	3.4568	3.3239	3.2172	3.1294	2.9931	2.8502	2.6993	2.5383	2.3637	2.1694
26	7.7213	5.5263	4.6366	4.1400	3.8183	3.5911	3.4210	3.2884	3.1818	3.0941	2.9579	2.8150	2.6640	2.5026	2.3273	2.1315
27	7.6767	5.4881	4.6009	4.1056	3.7848	3.5580	3.3882	3.2558	3.1494	3.0618	2.9256	2.7827	2.6316	2.4699	2.2938	2.0965
28	7.6356	5.4529	4.5681	4.0740	3.7539	3.5276	3.3581	3.2259	3.1195	3.0320	2.8959	2.7530	2.6017	2.4397	2.2629	2.0642
29	7.5976	5.4205	4.5378	4.0449	3.7254	3.4995	3.3302	3.1982	3.0920	3.0045	2.8685	2.7256	2.5742	2.4118	2.2344	2.0342
30	7.5625	5.3904	4.5097	4.0179	3.6990	3.4735	3.3045	3.1726	3.0665	2.9791	2.8431	2.7002	2.5487	2.3860	2.2079	2.0062
40	7.3141	5.1785	4.3126	3.8283	3.5138	3.2910	3.1238	2.9930	2.8876	2.8005	2.6648	2.5216	2.3689	2.2034	2.0194	1.8047
60	7.0771	4.9774	4.1259	3.6491	3.3389	3.1187	2.9530	2.8233	2.7185	2.6318	2.4961	2.3523	2.1978	2.0285	1.8363	1.6006
120	6.8510	4.7865	3.9493	3.4796	3.1735	2.9559	2.7918	2.6629	2.5586	2.4721	2.3363	2.1915	2.0346	1.8600	1.6557	1.3805
∞	6.6349	4.6052	3.7816	3.3192	3.0173	2.8020	2.6393	2.5113	2.4073	2.3209	2.1848	2.0385	1.8783	1.6964	1.4730	1.0000

Degrees of Freedom of the Numerator

Degrees of Freedom

F Distribution: upper 5 per cent points

Degrees of Freedom of the Numerator

$d.f._2$ \ $d.f._1$	1	2	3	4	5	6	7	8	9	10	12	15	20	30	60	∞
1	161·45	199·50	215·71	224·58	230·16	233·99	236·77	238·88	240·54	241·88	243·91	245·95	248·01	250·09	252·20	254·32
2	18·513	19·000	19·164	19·247	19·296	19·330	19·353	19·371	19·385	19·396	19·413	19·429	19·446	19·462	19·479	19·496
3	10·128	9·5521	9·2766	9·1172	9·0135	8·9406	8·8868	8·8452	8·8123	8·7855	8·7446	8·7029	8·6602	8·6166	8·5720	8·5265
4	7·7086	6·9443	6·5914	6·3883	6·2560	6·1631	6·0942	6·0410	5·9988	5·9644	5·9117	5·8578	5·8025	5·7459	5·6878	5·6281
5	6·6079	5·7861	5·4095	5·1922	5·0503	4·9503	4·8759	4·8183	4·7725	4·7351	4·6777	4·6188	4·5581	4·4957	4·4314	4·3650
6	5·9874	5·1433	4·7571	4·5337	4·3874	4·2839	4·2066	4·1468	4·0990	4·0600	3·9999	3·9381	3·8742	3·8082	3·7398	3·6688
7	5·5914	4·7374	4·3468	4·1203	3·9715	3·8660	3·7870	3·7257	3·6767	3·6365	3·5747	3·5108	3·4445	3·3758	3·3043	3·2298
8	5·3177	4·4590	4·0662	3·8378	3·6875	3·5806	3·5005	3·4381	3·3881	3·3472	3·2840	3·2184	3·1503	3·0794	3·0053	2·9276
9	5·1174	4·2565	3·8626	3·6331	3·4817	3·3738	3·2927	3·2296	3·1789	3·1373	3·0729	3·0061	2·9365	2·8637	2·7872	2·7067
10	4·9646	4·1028	3·7083	3·4780	3·3258	3·2172	3·1355	3·0717	3·0204	2·9782	2·9130	2·8450	2·7740	2·6996	2·6211	2·5379
11	4·8443	3·9823	3·5874	3·3567	3·2039	3·0946	3·0123	2·9480	2·8962	2·8536	2·7876	2·7186	2·6464	2·5705	2·4901	2·4045
12	4·7472	3·8853	3·4903	3·2592	3·1059	2·9961	2·9134	2·8486	2·7964	2·7534	2·6866	2·6169	2·5436	2·4663	2·3842	2·2962
13	4·6672	3·8056	3·4105	3·1791	3·0254	2·9153	2·8321	2·7669	2·7144	2·6710	2·6037	2·5331	2·4589	2·3803	2·2966	2·2064
14	4·6001	3·7389	3·3439	3·1122	2·9582	2·8477	2·7642	2·6987	2·6458	2·6021	2·5342	2·4630	2·3879	2·3082	2·2230	2·1307
15	4·5431	3·6823	3·2874	3·0556	2·9013	2·7905	2·7066	2·6408	2·5876	2·5437	2·4753	2·4035	2·3275	2·2468	2·1601	2·0658
16	4·4940	3·6337	3·2389	3·0069	2·8524	2·7413	2·6572	2·5911	2·5377	2·4935	2·4247	2·3522	2·2756	2·1938	2·1058	2·0096
17	4·4513	3·5915	3·1968	2·9647	2·8100	2·6987	2·6143	2·5480	2·4943	2·4499	2·3807	2·3077	2·2304	2·1477	2·0584	1·9604
18	4·4139	3·5546	3·1599	2·9277	2·7729	2·6613	2·5767	2·5102	2·4563	2·4117	2·3421	2·2686	2·1906	2·1071	2·0166	1·9168
19	4·3808	3·5219	3·1274	2·8951	2·7401	2·6283	2·5435	2·4768	2·4227	2·3779	2·3080	2·2341	2·1555	2·0712	1·9796	1·8780
20	4·3513	3·4928	3·0984	2·8661	2·7109	2·5990	2·5140	2·4471	2·3928	2·3479	2·2776	2·2033	2·1242	2·0391	1·9464	1·8432
21	4·3248	3·4668	3·0725	2·8401	2·6848	2·5727	2·4876	2·4205	2·3661	2·3210	2·2504	2·1757	2·0960	2·0102	1·9165	1·8117
22	4·3009	3·4434	3·0491	2·8167	2·6613	2·5491	2·4638	2·3965	2·3419	2·2967	2·2258	2·1508	2·0707	1·9842	1·8895	1·7831
23	4·2793	3·4221	3·0280	2·7955	2·6400	2·5277	2·4422	2·3748	2·3201	2·2747	2·2036	2·1282	2·0476	1·9605	1·8649	1·7570
24	4·2597	3·4028	3·0088	2·7763	2·6207	2·5082	2·4226	2·3551	2·3002	2·2547	2·1834	2·1077	2·0267	1·9390	1·8424	1·7331
25	4·2417	3·3852	2·9912	2·7587	2·6030	2·4904	2·4047	2·3371	2·2821	2·2365	2·1649	2·0889	2·0075	1·9192	1·8217	1·7110
26	4·2252	3·3690	2·9751	2·7426	2·5868	2·4741	2·3883	2·3205	2·2655	2·2197	2·1479	2·0716	1·9898	1·9010	1·8027	1·6906
27	4·2100	3·3541	2·9604	2·7278	2·5719	2·4591	2·3732	2·3053	2·2501	2·2043	2·1323	2·0558	1·9736	1·8842	1·7851	1·6717
28	4·1960	3·3404	2·9467	2·7141	2·5581	2·4453	2·3593	2·2913	2·2360	2·1900	2·1179	2·0411	1·9586	1·8687	1·7689	1·6541
29	4·1830	3·3277	2·9340	2·7014	2·5454	2·4324	2·3463	2·2782	2·2229	2·1768	2·1045	2·0275	1·9446	1·8543	1·7537	1·6377
30	4·1709	3·3158	2·9223	2·6896	2·5336	2·4205	2·3343	2·2662	2·2107	2·1646	2·0921	2·0148	1·9317	1·8409	1·7396	1·6223
40	4·0848	3·2317	2·8387	2·6060	2·4495	2·3359	2·2490	2·1802	2·1240	2·0772	2·0035	1·9245	1·8389	1·7444	1·6373	1·5089
60	4·0012	3·1504	2·7581	2·5252	2·3683	2·2540	2·1665	2·0970	2·0401	1·9926	1·9174	1·8364	1·7480	1·6491	1·5343	1·3893
120	3·9201	3·0718	2·6802	2·4472	2·2900	2·1750	2·0867	2·0164	1·9588	1·9105	1·8337	1·7505	1·6587	1·5543	1·4290	1·2539
∞	3·8415	2·9957	2·6049	2·3719	2·2141	2·0986	2·0096	1·9384	1·8799	1·8307	1·7522	1·6664	1·5705	1·4591	1·3180	1·0000

Degrees of Freedom of the Denominator

Reproduced with permission from Biometrika Tables for Statisticians

Use of the statistical tables

Normal probability table (Appendix D)
The table may be used to determine the probability that a single element in a set has a value above, below or between particular values. For example, given a set of normally distributed values, with $\bar{x} = 64\cdot71$ and $\sigma = 12\cdot34$, the probability that x_i falls between 70 and 80 is determined by calculating standardised values $z = \dfrac{x - \bar{x}}{\sigma}$. The two limits are given by

$$z_1 = \frac{70 - 64\cdot71}{12\cdot34} = 0\cdot43 \text{ and } z_2 = \frac{80 - 64\cdot71}{12\cdot34} = 1\cdot24.$$

The area to the left of $z_1 = 0\cdot43$ is $P_1 = 0\cdot6664$ and that to the left of $z_2 = 1\cdot24$ is $P_2 = 0\cdot8925$. The difference is $0\cdot8925 - 0\cdot6664 = 0\cdot2261$, and the probability that x_i lies between 70 and 80 is about $0\cdot226$.

At $z_1 = 0\cdot43$, the height of the ordinate,

$$y_1 = \frac{0\cdot3653 + 0\cdot3621}{2} = \frac{0\cdot7274}{2} = 0\cdot3637,$$

and at $z_2 = 1\cdot24$, $y_2 = 0\cdot1849$.

A distribution may be normalised by constructing the best-fitting normal curve which has the same mean and standard deviation as the original data. For each class of the data, calculate a standardised value $z = \dfrac{x_{\max} - \bar{x}}{s}$, where $x_{\max}$ is the upper limit of the class. From the table, find the proportion of the area of the normal curve (P) which lies below each value of z. Multiply the difference between successive values of P by the total number of items (n), to find the expected frequency (f_E). For example, if $n = 150$, $\bar{x} = 60\cdot6$, $s = 11\cdot8$ and the two highest values of $x_{\max}$ are 90 and 80, then the highest standardised value is $z_1 = \dfrac{90 - 60\cdot6}{11\cdot8} = \dfrac{29\cdot4}{11\cdot8} = 2\cdot49$. From the table, it can be seen that $0\cdot99361$ of the area of the normal curve (P_1) lies below this value. Similarly, for the second value, $z_2 = \dfrac{80 - 60\cdot6}{11\cdot8} = 1\cdot64$ and $P_2 = 0\cdot9495$. Then $P_1 - P_2 = 0\cdot0441$ and the expected frequency is $f_E = n(P_1 - P_2) = 6\cdot62$. The complete set of values of f_E may be compared with observed frequencies by means of the χ^2 test, and the goodness of fit can be determined.

Percentage points of the t distribution (Appendix F)
The table shows the percentage of the total area under the normal curve which is bounded by given values of t. For example, with 1 d.f., 30 per cent of the area lies outside the range $t = 0$ to $1\cdot963$, and $0\cdot1$ per cent lies out-

side the range $t = 0$ to $636 \cdot 62$; the probabilities that values of $t = 1 \cdot 963$ and $t = 636 \cdot 62$ occur by chance thus are $0 \cdot 3$ and $0 \cdot 001$, respectively.

To test the null hypothesis of no significant difference, find the tabulated values of t which lie immediately above and below the calculated value for the appropriate degrees of freedom. For example, with 60 d.f. and $t = 2 \cdot 14$, the bounding values are $2 \cdot 000$ ($2Q = 5$) and $2 \cdot 390$ ($2Q = 2$). Thus, the probability that $t = 2 \cdot 14$ arises by chance when there are 60 d.f. lies between $0 \cdot 02$ and $0 \cdot 05$, and the difference tested is probably significant.

Percentage points of the χ^2 distribution (Appendix G)

The table shows the probability that values of χ^2 larger than the calculated one arise by chance. For example, with 18 d.f., the probability that values of χ^2 greater than $8 \cdot 50$ could arise by chance lies between $97 \cdot 5$ per cent ($\chi^2 = 8 \cdot 231$) and $95 \cdot 0$ per cent ($\chi^2 = 9 \cdot 390$), approximately $1/5$ of the interval from $97 \cdot 5$ per cent; the probability is about $0 \cdot 97$ and the difference tested is not shown to be significant.

In testing goodness of fit, i.e. the null hypothesis that the difference between the observed distribution and the expected (e.g. normal) one is the result of chance, the same procedure is followed. For example, if the mean, standard deviation and number of cases (20) were not changed when the data were normalised, 3 d.f. are lost and 17 d.f. remain. A calculated value of $\chi^2 = 34 \cdot 6$ then lies approximately mid-way between the 1 per cent ($33 \cdot 409$) and $0 \cdot 5$ per cent ($35 \cdot 718$) values. Since the χ^2 value is larger than the 1 per cent level, the null hypothesis is rejected: the data do not come from a normally-distributed population.

F distribution (Appendix H)

F ratios calculated in the analysis of variance may be tested at the 5 per cent and 1 per cent levels. The appropriate table is entered at the position coinciding with d.f.$_1$ for the larger calculated value of the variance or mean square and with d.f.$_2$ for the smaller value. For example, if d.f.$_1 = 2$, d.f.$_2 = 60$ and $F = \dfrac{10 \cdot 8}{4 \cdot 6} = 2 \cdot 35$, the null hypothesis of no significant difference between the means of the sampled populations is not rejected, since the significant values at the 5 per cent and 1 per cent levels are $3 \cdot 1504$ and $4 \cdot 9774$ respectively.

If samples yield an F ratio of $\dfrac{5 \cdot 11}{1 \cdot 39} = 3 \cdot 68$ for d.f.$_1 = 7$ and d.f.$_2 = 40$, there is less than one chance in 100 that they were drawn from the same population, since the significant F values are $2 \cdot 2490$ (5 per cent) and $3 \cdot 1238$ (1 per cent).

Logarithms

	0	1	2	3	4	5	6	7	8	9	1	2	3	4	5	6	7	8	9
10	0000	0043	0086	0128	0170	0212	0253	0294	0334	0374	4	8	12	17	21	25	29	33	37
11	0414	0453	0492	0531	0569	0607	0645	0682	0719	0755	4	8	11	15	19	23	26	30	34
12	0792	0828	0864	0899	0934	0969	1004	1038	1072	1106	3	7	10	14	17	21	24	28	31
13	1139	1173	1206	1239	1271	1303	1335	1367	1399	1430	3	6	10	13	16	19	23	26	29
14	1461	1492	1523	1553	1584	1614	1644	1673	1703	1732	3	6	9	12	15	18	21	24	27
15	1761	1790	1818	1847	1875	1903	1931	1959	1987	2014	3	6	8	11	14	17	20	22	25
16	2041	2068	2095	2122	2148	2175	2201	2227	2253	2279	3	5	8	11	13	16	18	21	24
17	2304	2330	2355	2380	2405	2430	2455	2480	2504	2529	2	5	7	10	12	15	17	20	22
18	2553	2577	2601	2625	2648	2672	2695	2718	2742	2765	2	5	7	9	12	14	16	19	21
19	2788	2810	2833	2856	2878	2900	2923	2945	2967	2989	2	4	7	9	11	13	16	18	20
20	3010	3032	3054	3075	3096	3118	3139	3160	3181	3201	2	4	6	8	11	13	15	17	19
21	3222	3243	3263	3284	3304	3324	3345	3365	3385	3404	2	4	6	8	10	12	14	16	18
22	3424	3444	3464	3483	3502	3522	3541	3560	3579	3598	2	4	6	8	10	12	14	15	17
23	3617	3636	3655	3674	3692	3711	3729	3747	3766	3784	2	4	6	7	9	11	13	15	17
24	3802	3820	3838	3856	3874	3892	3909	3927	3945	3962	2	4	5	7	9	11	12	14	16
25	3979	3997	4014	4031	4048	4065	4082	4099	4116	4133	2	3	5	7	9	10	12	14	15
26	4150	4166	4183	4200	4216	4232	4249	4265	4281	4298	2	3	5	7	8	10	11	13	15
27	4314	4330	4346	4362	4378	4393	4409	4425	4440	4456	2	3	5	6	8	9	11	13	14
28	4472	4487	4502	4518	4533	4548	4564	4579	4594	4609	2	3	5	6	8	9	11	12	14
29	4624	4639	4654	4669	4683	4698	4713	4728	4742	4757	1	3	4	6	7	9	10	12	13
30	4771	4786	4800	4814	4829	4843	4857	4871	4886	4900	1	3	4	6	7	9	10	11	13
31	4914	4928	4942	4955	4969	4983	4997	5011	5024	5038	1	3	4	6	7	8	10	11	12
32	5051	5065	5079	5092	5105	5119	5132	5145	5159	5172	1	3	4	5	7	8	9	11	12
33	5185	5198	5211	5224	5237	5250	5263	5276	5289	5302	1	3	4	5	6	8	9	10	12
34	5315	5328	5340	5353	5366	5378	5391	5403	5416	5428	1	3	4	5	6	8	9	10	11
35	5441	5453	5465	5478	5490	5502	5514	5527	5539	5551	1	2	4	5	6	7	9	10	11
36	5563	5575	5587	5599	5611	5623	5635	5647	5658	5670	1	2	4	5	6	7	8	10	11
37	5682	5694	5705	5717	5729	5740	5752	5763	5775	5786	1	2	3	5	6	7	8	9	10
38	5798	5809	5821	5832	5843	5855	5866	5877	5888	5899	1	2	3	5	6	7	8	9	10
39	5911	5922	5933	5944	5955	5966	5977	5988	5999	6010	1	2	3	4	5	7	8	9	10
40	6021	6031	6042	6053	6064	6075	6085	6096	6107	6117	1	2	3	4	5	6	8	9	10
41	6128	6138	6149	6160	6170	6180	6191	6201	6212	6222	1	2	3	4	5	6	7	8	9
42	6232	6243	6253	6263	6274	6284	6294	6304	6314	6325	1	2	3	4	5	6	7	8	9
43	6335	6345	6355	6365	6375	6385	6395	6405	6415	6425	1	2	3	4	5	6	7	8	9
44	6435	6444	6454	6464	6474	6484	6493	6503	6513	6522	1	2	3	4	5	6	7	8	9
45	6532	6542	6551	6561	6571	6580	6590	6599	6609	6618	1	2	3	4	5	6	7	8	9
46	6628	6637	6646	6656	6665	6675	6684	6693	6702	6712	1	2	3	4	5	6	7	7	8
47	6721	6730	6739	6749	6758	6767	6776	6785	6794	6803	1	2	3	4	5	5	6	7	8
48	6812	6821	6830	6839	6848	6857	6866	6875	6884	6893	1	2	3	4	4	5	6	7	8
49	6902	6911	6920	6928	6937	6946	6955	6964	6972	6981	1	2	3	4	4	5	6	7	8
50	6990	6998	7007	7016	7024	7033	7042	7050	7059	7067	1	2	3	3	4	5	6	7	8
51	7076	7084	7093	7101	7110	7118	7126	7135	7143	7152	1	2	3	3	4	5	6	7	8
52	7160	7168	7177	7185	7193	7202	7210	7218	7226	7235	1	2	2	3	4	5	6	7	7
53	7243	7251	7259	7267	7275	7284	7292	7300	7308	7316	1	2	2	3	4	5	6	6	7
54	7324	7332	7340	7348	7356	7364	7372	7380	7388	7396	1	2	2	3	4	5	6	6	7
	0	1	2	3	4	5	6	7	8	9	1	2	3	4	5	6	7	8	9

	0	1	2	3	4	5	6	7	8	9	1	2	3	4	5	6	7	8	9
55	7404	7412	7419	7427	7435	7443	7451	7459	7466	7474	1	2	2	3	4	5	5	6	7
56	7482	7490	7497	7505	7513	7520	7528	7536	7543	7551	1	2	2	3	4	5	5	6	7
57	7559	7566	7574	7582	7589	7597	7604	7612	7619	7627	1	2	2	3	4	5	5	6	7
58	7634	7642	7649	7657	7664	7672	7679	7686	7694	7701	1	1	2	3	4	4	5	6	7
59	7709	7716	7723	7731	7738	7745	7752	7760	7767	7774	1	1	2	3	4	4	5	6	7
60	7782	7789	7796	7803	7810	7818	7825	7832	7839	7846	1	1	2	3	4	4	5	6	6
61	7853	7860	7868	7875	7882	7889	7896	7903	7910	7917	1	1	2	3	4	4	5	6	6
62	7924	7931	7938	7945	7952	7959	7966	7973	7980	7987	1	1	2	3	3	4	5	6	6
63	7993	8000	8007	8014	8021	8028	8035	8041	8048	8055	1	1	2	3	3	4	5	5	6
64	8062	8069	8075	8082	8089	8096	8102	8109	8116	8122	1	1	2	3	3	4	5	5	6
65	8129	8136	8142	8149	8156	8162	8169	8176	8182	8189	1	1	2	3	3	4	5	5	6
66	8195	8202	8209	8215	8222	8228	8235	8241	8248	8254	1	1	2	3	3	4	5	5	6
67	8261	8267	8274	8280	8287	8293	8299	8306	8312	8319	1	1	2	3	3	4	5	5	6
68	8325	8331	8338	8344	8351	8357	8363	8370	8376	8382	1	1	2	3	3	4	4	5	6
69	8388	8395	8401	8407	8414	8420	8426	8432	8439	8445	1	1	2	2	3	4	4	5	6
70	8451	8457	8463	8470	8476	8482	8488	8494	8500	8506	1	1	2	2	3	4	4	5	6
71	8513	8519	8525	8531	8537	8543	8549	8555	8561	8567	1	1	2	2	3	4	4	5	5
72	8573	8579	8585	8591	8597	8603	8609	8615	8621	8627	1	1	2	2	3	4	4	5	5
73	8633	8639	8645	8651	8657	8663	8669	8675	8681	8686	1	1	2	2	3	4	4	5	5
74	8692	8698	8704	8710	8716	8722	8727	8733	8739	8745	1	1	2	2	3	4	4	5	5
75	8751	8756	8762	8768	8774	8779	8785	8791	8797	8802	1	1	2	2	3	3	4	5	5
76	8808	8814	8820	8825	8831	8837	8842	8848	8854	8859	1	1	2	2	3	3	4	5	5
77	8865	8871	8876	8882	8887	8893	8899	8904	8910	8915	1	1	2	2	3	3	4	4	5
78	8921	8927	8932	8938	8943	8949	8954	8960	8965	8971	1	1	2	2	3	3	4	4	5
79	8976	8982	8987	8993	8998	9004	9009	9015	9020	9025	1	1	2	2	3	3	4	4	5
80	9031	9036	9042	9047	9053	9058	9063	9069	9074	9079	1	1	2	2	3	3	4	4	5
81	9085	9090	9096	9101	9106	9112	9117	9122	9128	9133	1	1	2	2	3	3	4	4	5
82	9138	9143	9149	9154	9159	9165	9170	9175	9180	9186	1	1	2	2	3	3	4	4	5
83	9191	9196	9201	9206	9212	9217	9222	9227	9232	9238	1	1	2	2	3	3	4	4	5
84	9243	9248	9253	9258	9263	9269	9274	9279	9284	9289	1	1	2	2	3	3	4	4	5
85	9294	9299	9304	9309	9315	9320	9325	9330	9335	9340	1	1	2	2	3	3	4	4	5
86	9345	9350	9355	9360	9365	9370	9375	9380	9385	9390	1	1	2	2	3	3	4	4	5
87	9395	9400	9405	9410	9415	9420	9425	9430	9435	9440	0	1	1	2	2	3	3	4	4
88	9445	9450	9455	9460	9465	9469	9474	9479	9484	9489	0	1	1	2	2	3	3	4	4
89	9494	9499	9504	9509	9513	9518	9523	9528	9533	9538	0	1	1	2	2	3	3	4	4
90	9542	9547	9552	9557	9562	9566	9571	9576	9581	9586	0	1	1	2	2	3	3	4	4
91	9590	9595	9600	9605	9609	9614	9619	9624	9628	9633	0	1	1	2	2	3	3	4	4
92	9638	9643	9647	9652	9657	9661	9666	9671	9675	9680	0	1	1	2	2	3	3	4	4
93	9685	9689	9694	9699	9703	9708	9713	9717	9722	9727	0	1	1	2	2	3	3	4	4
94	9731	9736	9741	9745	9750	9754	9759	9763	9768	9773	0	1	1	2	2	3	3	4	4
95	9777	9782	9786	9791	9795	9800	9805	9809	9814	9818	0	1	1	2	2	3	3	4	4
96	9823	9827	9832	9836	9841	9845	9850	9854	9859	9863	0	1	1	2	2	3	3	4	4
97	9868	9872	9877	9881	9886	9890	9894	9899	9903	9908	0	1	1	2	2	3	3	4	4
98	9912	9917	9921	9926	9930	9934	9939	9943	9948	9952	0	1	1	2	2	3	3	4	4
99	9956	9961	9965	9969	9974	9978	9983	9987	9991	9996	0	1	1	2	2	3	3	3	4
	0	1	2	3	4	5	6	7	8	9	1	2	3	4	5	6	7	8	9

Antilogarithms

	0	1	2	3	4	5	6	7	8	9	1	2	3	4	5	6	7	8	9
·00	1000	1002	1005	1007	1009	1012	1014	1016	1019	1021	0	0	1	1	1	1	2	2	2
·01	1023	1026	1028	1030	1033	1035	1038	1040	1042	1045	0	0	1	1	1	1	2	2	2
·02	1047	1050	1052	1054	1057	1059	1062	1064	1067	1069	0	0	1	1	1	1	2	2	2
·03	1072	1074	1076	1079	1081	1084	1086	1089	1091	1094	0	0	1	1	1	1	2	2	2
·04	1096	1099	1102	1104	1107	1109	1112	1114	1117	1119	0	1	1	1	1	2	2	2	2
·05	1122	1125	1127	1130	1132	1135	1138	1140	1143	1146	0	1	1	1	1	2	2	2	2
·06	1148	1151	1153	1156	1159	1161	1164	1167	1169	1172	0	1	1	1	1	2	2	2	2
·07	1175	1178	1180	1183	1186	1189	1191	1194	1197	1199	0	1	1	1	1	2	2	2	2
·08	1202	1205	1208	1211	1213	1216	1219	1222	1225	1227	0	1	1	1	1	2	2	2	3
·09	1230	1233	1236	1239	1242	1245	1247	1250	1253	1256	0	1	1	1	1	2	2	2	3
·10	1259	1262	1265	1268	1271	1274	1276	1279	1282	1285	0	1	1	1	1	2	2	2	3
·11	1288	1291	1294	1297	1300	1303	1306	1309	1312	1315	0	1	1	1	2	2	2	2	3
·12	1318	1321	1324	1327	1330	1334	1337	1340	1343	1346	0	1	1	1	2	2	2	2	3
·13	1349	1352	1355	1358	1361	1365	1368	1371	1374	1377	0	1	1	1	2	2	2	3	3
·14	1380	1384	1387	1390	1393	1396	1400	1403	1406	1409	0	1	1	1	2	2	2	3	3
·15	1413	1416	1419	1422	1426	1429	1432	1435	1439	1442	0	1	1	1	2	2	2	3	3
·16	1445	1449	1452	1455	1459	1462	1466	1469	1472	1476	0	1	1	1	2	2	2	3	3
·17	1479	1483	1486	1489	1493	1496	1500	1503	1507	1510	0	1	1	1	2	2	2	3	3
·18	1514	1517	1521	1524	1528	1531	1535	1538	1542	1545	0	1	1	1	2	2	2	3	3
·19	1549	1552	1556	1560	1563	1567	1570	1574	1578	1581	0	1	1	1	2	2	3	3	3
·20	1585	1589	1592	1596	1600	1603	1607	1611	1614	1618	0	1	1	1	2	2	3	3	3
·21	1622	1626	1629	1633	1637	1641	1644	1648	1652	1656	0	1	1	2	2	2	3	3	3
·22	1660	1663	1667	1671	1675	1679	1683	1687	1690	1694	0	1	1	2	2	2	3	3	3
·23	1698	1702	1706	1710	1714	1718	1722	1726	1730	1734	0	1	1	2	2	2	3	3	4
·24	1738	1742	1746	1750	1754	1758	1762	1766	1770	1774	0	1	1	2	2	2	3	3	4
·25	1778	1782	1786	1791	1795	1799	1803	1807	1811	1816	0	1	1	2	2	2	3	3	4
·26	1820	1824	1828	1832	1837	1841	1845	1849	1854	1858	0	1	1	2	2	3	3	3	4
·27	1862	1866	1871	1875	1879	1884	1888	1892	1897	1901	0	1	1	2	2	3	3	3	4
·28	1905	1910	1914	1919	1923	1928	1932	1936	1941	1945	0	1	1	2	2	3	3	4	4
·29	1950	1954	1959	1963	1968	1972	1977	1982	1986	1991	0	1	1	2	2	3	3	4	4
·30	1995	2000	2004	2009	2014	2018	2023	2028	2032	2037	0	1	1	2	2	3	3	4	4
·31	2042	2046	2051	2056	2061	2065	2070	2075	2080	2084	0	1	1	2	2	3	3	4	4
·32	2089	2094	2099	2104	2109	2113	2118	2123	2128	2133	0	1	1	2	2	3	3	4	4
·33	2138	2143	2148	2153	2158	2163	2168	2173	2178	2183	0	1	1	2	2	3	3	4	4
·34	2188	2193	2198	2203	2208	2213	2218	2223	2228	2234	1	1	2	2	3	3	4	4	5
·35	2239	2244	2249	2254	2259	2265	2270	2275	2280	2286	1	1	2	2	3	3	4	4	5
·36	2291	2296	2301	2307	2312	2317	2323	2328	2333	2339	1	1	2	2	3	3	4	4	5
·37	2344	2350	2355	2360	2366	2371	2377	2382	2388	2393	1	1	2	2	3	3	4	4	5
·38	2399	2404	2410	2415	2421	2427	2432	2438	2443	2449	1	1	2	2	3	3	4	4	5
·39	2455	2460	2466	2472	2477	2483	2489	2495	2500	2506	1	1	2	2	3	3	4	5	5
·40	2512	2518	2523	2529	2535	2541	2547	2553	2559	2564	1	1	2	2	3	4	4	5	5
·41	2570	2576	2582	2588	2594	2600	2606	2612	2618	2624	1	1	2	2	3	4	4	5	5
·42	2630	2636	2642	2649	2655	2661	2667	2673	2679	2685	1	1	2	2	3	4	4	5	6
·43	2692	2698	2704	2710	2716	2723	2729	2735	2742	2748	1	1	2	3	3	4	4	5	6
·44	2754	2761	2767	2773	2780	2786	2793	2799	2805	2812	1	1	2	3	3	4	4	5	6
·45	2818	2825	2831	2838	2844	2851	2858	2864	2871	2877	1	1	2	3	3	4	5	5	6
·46	2884	2891	2897	2904	2911	2917	2924	2931	2938	2944	1	1	2	3	3	4	5	5	6
·47	2951	2958	2965	2972	2979	2985	2992	2999	3006	3013	1	1	2	3	3	4	5	5	6
·48	3020	3027	3034	3041	3048	3055	3062	3069	3076	3083	1	1	2	3	4	4	5	6	6
·49	3090	3097	3105	3112	3119	3126	3133	3141	3148	3155	1	1	2	3	4	4	5	6	6
	0	1	2	3	4	5	6	7	8	9	1	2	3	4	5	6	7	8	9

	0	1	2	3	4	5	6	7	8	9	1	2	3	4	5	6	7	8	9
50	3162	3170	3177	3184	3192	3199	3206	3214	3221	3228	1	1	2	3	4	4	5	6	7
51	3236	3243	3251	3258	3266	3273	3281	3289	3296	3304	1	2	2	3	4	5	5	6	7
52	3311	3319	3327	3334	3342	3350	3357	3365	3373	3381	1	2	2	3	4	5	5	6	7
53	3388	3396	3404	3412	3420	3428	3436	3443	3451	3459	1	2	2	3	4	5	6	6	7
54	3467	3475	3483	3491	3499	3508	3516	3524	3532	3540	1	2	2	3	4	5	6	6	7
55	3548	3556	3565	3573	3581	3589	3597	3606	3614	3622	1	2	2	3	4	5	6	7	7
56	3631	3639	3648	3656	3664	3673	3681	3690	3698	3707	1	2	3	3	4	5	6	7	8
57	3715	3724	3733	3741	3750	3758	3767	3776	3784	3793	1	2	3	3	4	5	6	7	8
58	3802	3811	3819	3828	3837	3846	3855	3864	3873	3882	1	2	3	4	4	5	6	7	8
59	3890	3899	3908	3917	3926	3936	3945	3954	3963	3972	1	2	3	4	5	5	6	7	8
60	3981	3990	3999	4009	4018	4027	4036	4046	4055	4064	1	2	3	4	5	6	6	7	8
61	4074	4083	4093	4102	4111	4121	4130	4140	4150	4159	1	2	3	4	5	6	7	8	9
62	4169	4178	4188	4198	4207	4217	4227	4236	4246	4256	1	2	3	4	5	6	7	8	9
63	4266	4276	4285	4295	4305	4315	4325	4335	4345	4355	1	2	3	4	5	6	7	8	9
64	4365	4375	4385	4395	4406	4416	4426	4436	4446	4457	1	2	3	4	5	6	7	8	9
65	4467	4477	4487	4498	4508	4519	4529	4539	4550	4560	1	2	3	4	5	6	7	8	9
66	4571	4581	4592	4603	4613	4624	4634	4645	4656	4667	1	2	3	4	5	6	7	9	10
67	4677	4688	4699	4710	4721	4732	4742	4753	4764	4775	1	2	3	4	5	7	8	9	10
68	4786	4797	4808	4819	4831	4842	4853	4864	4875	4887	1	2	3	4	6	7	8	9	10
69	4898	4909	4920	4932	4943	4955	4966	4977	4989	5000	1	2	3	5	6	7	8	9	10
70	5012	5023	5035	5047	5058	5070	5082	5093	5105	5117	1	2	4	5	6	7	8	9	11
71	5129	5140	5152	5164	5176	5188	5200	5212	5224	5236	1	2	4	5	6	7	8	10	11
72	5248	5260	5272	5284	5297	5309	5321	5333	5346	5358	1	2	4	5	6	7	9	10	11
73	5370	5383	5395	5408	5420	5433	5445	5458	5470	5483	1	3	4	5	6	8	9	10	11
74	5495	5508	5521	5534	5546	5559	5572	5585	5598	5610	1	3	4	5	6	8	9	10	12
75	5623	5636	5649	5662	5675	5689	5702	5715	5728	5741	1	3	4	5	7	8	9	10	12
76	5754	5768	5781	5794	5808	5821	5834	5848	5861	5875	1	3	4	5	7	8	9	11	12
77	5888	5902	5916	5929	5943	5957	5970	5984	5998	6012	1	3	4	5	7	8	10	11	12
78	6026	6039	6053	6067	6081	6095	6109	6124	6138	6152	1	3	4	6	7	8	10	11	13
79	6166	6180	6194	6209	6223	6237	6252	6266	6281	6295	1	3	4	6	7	9	10	11	13
80	6310	6324	6339	6353	6368	6383	6397	6412	6427	6442	1	3	4	6	7	9	10	12	13
81	6457	6471	6486	6501	6516	6531	6546	6561	6577	6592	2	3	5	6	8	9	11	12	14
82	6607	6622	6637	6653	6668	6683	6699	6714	6730	6745	2	3	5	6	8	9	11	12	14
83	6761	6776	6792	6808	6823	6839	6855	6871	6887	6902	2	3	5	6	8	9	11	13	14
84	6918	6934	6950	6966	6982	6998	7015	7031	7047	7063	2	3	5	6	8	10	11	13	15
85	7079	7096	7112	7129	7145	7161	7178	7194	7211	7228	2	3	5	7	8	10	12	13	15
86	7244	7261	7278	7295	7311	7328	7345	7362	7379	7396	2	3	5	7	8	10	12	13	15
87	7413	7430	7447	7464	7482	7499	7516	7534	7551	7568	2	3	5	7	9	10	12	14	16
88	7586	7603	7621	7638	7656	7674	7691	7709	7727	7745	2	4	5	7	9	11	12	14	16
89	7762	7780	7798	7816	7834	7852	7870	7889	7907	7925	2	4	5	7	9	11	13	14	16
90	7943	7962	7980	7998	8017	8035	8054	8072	8091	8110	2	4	6	7	9	11	13	15	17
91	8128	8147	8166	8185	8204	8222	8241	8260	8279	8299	2	4	6	8	9	11	13	15	17
92	8318	8337	8356	8375	8395	8414	8433	8453	8472	8492	2	4	6	8	10	12	14	15	17
93	8511	8531	8551	8570	8590	8610	8630	8650	8670	8690	2	4	6	8	10	12	14	16	18
94	8710	8730	8750	8770	8790	8810	8831	8851	8872	8892	2	4	6	8	10	12	14	16	18
95	8913	8933	8954	8974	8995	9016	9036	9057	9078	9099	2	4	6	8	10	12	15	17	19
96	9120	9141	9162	9183	9204	9226	9247	9268	9290	9311	2	4	6	8	11	13	15	17	19
97	9333	9354	9376	9397	9419	9441	9462	9484	9506	9528	2	4	7	9	11	13	15	17	20
98	9550	9572	9594	9616	9638	9661	9683	9705	9727	9750	2	4	7	9	11	13	16	18	20
99	9772	9795	9817	9840	9863	9836	9908	9931	9954	9977	2	5	7	9	11	14	16	18	20
	0	1	2	3	4	5	6	7	8	9	1	2	3	4	5	6	7	8	9

Glossary of symbols

A Area

a Regression coefficient

b Intercept of a regression line on an axis

c Class interval

d Difference or deviation from a given value

$\bar{d}$ Mean distance between points

d.f. Degrees of freedom

E Expected value

f Frequency or sampling fraction

f_c Cell frequency

k Bessel's correction

L.Q. Location quotient

N, n Number. (Where a population and sample are being compared, n is the number in the sample and N the number in the population.)

O Observed value

P_i Population of the ith point or area

p, q Probabilities (where $p + q = 1$)

R_n Nearest neighbour index

R_{xy} Product-moment coefficient of correlation

s Sample standard deviation

s^2 Sample variance

S.D. Standard distance

S.E. Standard error

T Correction factor for tied values in calculation of coefficient of rank correlation

t_c Critical value (Student's t distribution)

V Coefficient of variation

$\bar{X}$ Population mean

$\bar{X}_m$ Population median

$|x|$ x without reference to sign

x_i Any one unit of x

x_0 Mid-class value

$x_1, x_2, \ldots, x_n$ Individual units of x (total n)

$\bar{x}$ Sample mean

$\bar{x}_m$ Sample median

$\bar{x}_0$ Assumed mean or arbitrary origin

(x_i, y_i) Co-ordinates of the ith point

$(\bar{x}, \bar{y})$ Mean of an array of points

$(\bar{x}_c, \bar{y}_c)$ Co-ordinates of the mean centre

$(\bar{x}_m, \bar{y}_m)$ Co-ordinates of the median centre

z Units of the x axis on the normal curve
z_c Confidence coefficient
π Pi
ρ Coefficient of rank correlation
Σ Sum of
σ Population standard deviation
σ^2 Population variance
σ_p Standard error of a proportion
$\hat{\sigma}$ Best estimate of population standard deviation
σ_{xy} Covariance
χ^2 Chi squared

Glossary of terms

ANALYSIS OF VARIANCE: A technique which assesses the contribution made by each separate factor to the total variability of a set of data.

AVERAGE: Any measure of central tendency.

BESSEL'S CORRECTION (k): A factor $\sqrt{\dfrac{n}{n-1}}$ which converts the sample standard deviation into the best estimate of the population standard deviation.

BINOMIAL: A term used to describe a set which consists of two groups or classes of elements, a situation in which there are two possible outcomes of an event, etc.

BINOMIAL FORMULA: The expansion of the expression $(p + q)^n$, which has general application to the probability of outcomes of a binomial situation.

CENTRAL TENDENCY: The tendency of values of individual items within a set to cluster about a particular value or values, such as the arithmetic mean, median, etc.

CHI SQUARED (χ^2): $\chi^2 = \sum \left[\dfrac{(O - E)^2}{E} \right]$

CORRELATION COEFFICIENT: The numerical relationship between pairs of variables. (See Coefficient of Rank Correlation and Product–Moment Coefficient of Correlation.)

COEFFICIENT OF DETERMINATION: The percentage of the variation in one variable which is explained by variations in the other.
$(= R_{xy}^2 \times 100 \text{ per cent})$

COEFFICIENT OF RANK CORRELATION: The product-moment coefficient of correlation calculated from the ranks of two sets of variables, rather than from their absolute values.

COEFFICIENT OF VARIATION (V): $V = \dfrac{100\sigma}{\bar{x}}$ per cent

COLUMN: A set of numbers arranged vertically.

CONFIDENCE COEFFICIENT (or Critical value): The numerical values of confidence limits.

CONFIDENCE INTERVAL: A range of values within which the value of a parameter is expected to lie. The probability that the parameter does lie within the interval may be specified.

CONFIDENCE LEVEL: The probability that the value of a parameter lies within a particular confidence interval.

CONFIDENCE LIMIT: The end values of confidence intervals.

COVARIANCE: An average of the variation of variates in different groups.
$$\sigma_{xy} = \frac{\Sigma (x - \bar{x})(y - \bar{y})}{n}$$

DEGREES OF FREEDOM (d.f.): The maximum number of variates that can be assigned values before the others become determined.

ELEMENTS: The members in a set.

GAUSSIAN CURVE: (See Normal Probability Curve).

HISTOGRAM: A graph displaying the frequency of items within classes.

INTERQUARTILE RANGE: The difference between the 25th and 75th percentiles.

LAW OF ADDITION: Where categories are mutually exclusive, their individual probabilities of occurrence or selection may be added.

LAW OF MULTIPLICATION: If events are independent of each other, the probability that both will occur is the product of their individual probabilities.

LEAST SQUARES METHOD: A method of fitting a line to a scatter of points in such a way that, if the distance between each point and the line is squared, the total of all these squares is at a minimum.

LOCATION QUOTIENT (L.Q.): A measure which compares an area's share of national employment in an industry with its share of all national employment.

LOG-NORMAL DISTRIBUTION: A distribution which is normal when transformed by using the logarithms of the numbers in the scale instead of the numbers themselves.

LORENZ CURVE: A curve showing how a particular distribution compares with an even one.

MEAN (ARITHMETIC, $\bar{x}$): $\bar{x} = \dfrac{\Sigma x}{N}$

MEAN CENTRE ($\bar{x}_c, \bar{y}_c$): The centre of gravity of a distribution
$$\left(= \frac{\Sigma(x_i P_i)}{\Sigma P_i}, \frac{\Sigma(y_i P_i)}{\Sigma P_i} \right)$$

MEAN DEVIATION: The mean value of all individual deviations from a given value, without reference to sign
$$\left(= \frac{1}{N}\Sigma|d| \right)$$

MEAN OF ARRAY ($\bar{x}, \bar{y}$): The point ($\bar{x}, \bar{y}$) in a bivariate distribution.

MEDIAN: The central value in an ordered series, i.e. that value with an equal number of values above and below it. In a cumulative frequency curve it is the 50th percentile.

MODAL CLASS: The largest class in a frequency distribution. Distributions may be unimodal (with one modal class), bimodal (with two) etc.

MOMENT: A numerical quantity computed from a distribution. The order of the moment is determined by the power of the variable, i.e. the first moment (the mean) concerns x, the second concerns x^2, etc.

MOVING MEAN: (See Running Mean.)

NEAREST NEIGHBOUR INDEX (R_n): $R_n = 2\bar{d}\sqrt{\dfrac{n}{A}}$

NORMAL PROBABILITY CURVE (or Gaussian curve): A curve with the equation
$$y = \frac{1}{\sigma\sqrt{2\pi}} e^{\frac{-(x-\bar{x})^2}{2\sigma^2}}$$

NULL HYPOTHESIS: The assumption of no significant difference made when applying a significance test.

PARAMETER: Any statistical measure (e.g. mean, standard deviation).

PASCAL'S TRIANGLE: The coefficients of the expansion of $(p + q)^n$.

PERCENTILE: The value below which lies a particular percentage of a distribution.

POPULATION: Any set of items.

PROBABILITY PAPER: Graph paper on which a cumulative normal frequency curve appears as a straight line.

PRODUCT-MOMENT COEFFICIENT OF CORRELATION: The ratio of the covariance of two variables to the product of their standard deviations.

QUARTILES: The percentiles which divide a distribution into quarters (i.e. 25th, 50th or median and 75th percentiles). The 25th and 75th percentiles are known respectively as the lower and upper quartiles.

RANDOM SAMPLE: A sample drawn in such a way that every item in the population has an equal chance of selection.

RANKING: Arranging in order (of size, etc.).

REGRESSION LINE: A best-fitting line through a scatter of points.

REGRESSION COEFFICENT (a): The slope of the regression line.

RESIDUAL: In regression, the perpendicular distance between the line and a point.

ROW: A set of numbers arranged horizontally.

RUNNING MEAN: A smoothing device, designed to eliminate erratic or short-term movements in a time series. If the terms in the series are $x_1, x_2, \ldots, x_n$, the successive mean values $\dfrac{x_1 + x_2 + \ldots + x_i}{i}$,

$\dfrac{x_2 + x_3 + \ldots + x_{(i+1)}}{i}, \ldots \dfrac{x_{(n-i)} + x_{(n-i+1)} + \ldots + x_n}{i}$ are plotted

at the positions of the $\left(\dfrac{i+1}{2}\right)$th, $\left(\dfrac{(i+1)+2}{2}\right)$th $\ldots \left(\dfrac{n+(n-1)}{2}\right)$th terms in the series.

SAMPLE: A subset of a population.

SAMPLING DISTRIBUTION: The distribution of a particular statistic of all possible samples of a given size.

SAMPLE FRACTION (f): The proportion of a population forming a sample.

SET: A well-defined collection of objects.

SIGNIFICANCE: The probability that a null hypothesis is true.

SKEWNESS: The degree of asymmetry of a distribution, i.e. the extent to which the mean differs from the median.

STANDARD DEVIATION: A quantity describing the distribution of a set of items which is expressed in the same units as the items themselves. It indicates the degree to which individual values cluster around the mean and may be used as a measure of the variability of a frequency distribution.

$$\sigma = \sqrt{\frac{\Sigma(x - \bar{x})^2}{N}} \quad \text{or} \quad \sqrt{\frac{\Sigma x^2}{N} - \bar{x}^2}$$

STANDARD DISTANCE: The spatial equivalent of the standard deviation.

$$\text{S.D.} = \sqrt{\frac{\Sigma[P_i(x_i - \bar{x}_i)^2]}{\Sigma P_i} + \frac{\Sigma[P_i(y_i - \bar{y}_i)^2]}{\Sigma P_i}}$$

STATISTIC: The sample estimate of a parameter.

STRATIFIED SAMPLE: A sample in which several divisions of the population are sampled in proportion to their size.

SUBSET: Part of a set of objects.

SYSTEMATIC RANDOM SAMPLE: A random sample within a systematic framework.

SYSTEMATIC SAMPLE: A sample taken at regular intervals (space, time, etc.).

TOPOLOGY: Preservation of relationships (e.g. area) without concern for distance or direction.

TRANSFORMATION: The modification of data to make it more manageable.

VARIABLE: An item which can have several values.

VARIANCE: The square of the standard deviation.

VARIANCE RATIO: The relation of the larger of two variances to the smaller.

VARIATE: Any one value of a variable.

Index